Change in Global Environmental Politics

As wildfires rage, the cover of pollution thickens, and species disappear, the world confronts environmental crisis with a set of global institutions in urgent need of reform. Yet, these institutions have proved frustratingly resistant to change. Introducing the concept of Temporal Focal Points, Michael W. Manulak shows how change can – and does – occur in world politics. By reenvisioning the role of timing and temporality in social relations, his analysis presents a new approach to understanding transformative phases in international cooperation. We may now be entering such a phase, he argues, and global actors must be ready to realize the opportunities presented. Charting the often colorful and intensely political history of change in global environmental politics, this book sheds new light on the actors and institutions that shape humanity's response to planetary decline.

MICHAEL W. MANULAK is Assistant Professor at the Norman Paterson School of International Affairs, Carleton University. His research focuses on international organizations, global environmental politics, and international security. From 2015 to 2019 he served as an analyst in the Canadian government. He obtained his D.Phil. in international relations at St. Antony's College, University of Oxford.

Change in Global Environmental Politics

Temporal Focal Points and the Reform of International Institutions

MICHAEL W. MANULAK
Carleton University

CAMBRIDGE
UNIVERSITY PRESS

CAMBRIDGE
UNIVERSITY PRESS

University Printing House, Cambridge CB2 8BS, United Kingdom

One Liberty Plaza, 20th Floor, New York, NY 10006, USA

477 Williamstown Road, Port Melbourne, VIC 3207, Australia

314–321, 3rd Floor, Plot 3, Splendor Forum, Jasola District Centre, New Delhi – 110025, India

103 Penang Road, #05–06/07, Visioncrest Commercial, Singapore 238467

Cambridge University Press is part of the University of Cambridge.

It furthers the University's mission by disseminating knowledge in the pursuit of education, learning, and research at the highest international levels of excellence.

www.cambridge.org
Information on this title: www.cambridge.org/9781009165884
DOI: 10.1017/9781009165877

First published 2022

A catalogue record for this publication is available from the British Library.

Library of Congress Cataloging-in-Publication Data
Names: Manulak, Michael W., 1983- author.
Title: Change in global environmental politics : temporal focal points and the reform of international institutions / Michael W. Manulak, Carleton University, Ottawa.
Description: First Edition. | New York : Cambridge University Press, 2022. | Includes bibliographical references and index.
Identifiers: LCCN 2021058158 (print) | LCCN 2021058159 (ebook) | ISBN 9781009165884 (Hardback) | ISBN 9781009165891 (Paperback) | ISBN 9781009165877 (ePub)
Subjects: LCSH: Environmental policy–International cooperation. | Environmental management. | Environmental protection–Political aspects.
Classification: LCC GE170 .M368 2022 (print) | LCC GE170 (ebook) | DDC 363.7/0526–dc23/eng/20211214
LC record available at https://lccn.loc.gov/2021058158
LC ebook record available at https://lccn.loc.gov/2021058159

ISBN 978-1-009-16588-4 Hardback

To Jee, Theodore, and Sophia

Contents

Figures and Tables

Figures

Tables

Acknowledgments

This book would not have been possible without the support of a wide group of friends and colleagues. My greatest debt is to Duncan Snidal, a tremendous mentor, doctoral supervisor, and friend. He encouraged me to pursue the Temporal Focal Point framework when it was a fuzzy idea and challenged me to make it better. He guided and nurtured this project with generosity and graciousness, even long after I completed my degree. In the process, I became a political scientist. I am grateful to Norman Hillmer for his support and friendship. His empathy, advice, and incisive comments on multiple drafts of this book made the final product better and much more readable. I would like to thank Anne Deighton for her supervision in the early stages of my doctoral work, and Margaret MacMillan, a tremendous support and inspiration as I developed the historical side of this project, for her advice. While researching and writing the dissertation that formed the basis for this book, the University of Oxford's Department of Politics and International Relations, St. Antony's College, and Goodenough College in London provided tremendous intellectual environments.

In revising and completing this book, I've incurred huge intellectual debts to those who have read and commented on all, or some, of the book. I'd like to give special thanks to those who participated in my book workshop: Steven Bernstein, Orfeo Fioretos, Fen Hampson, Maria Ivanova, Phillip Lipscy, Louis Pauly, Stephen Saideman, Alexander Thompson, Inger Weibust, and David Welch. I am grateful to the Norman Paterson School of International Affairs (NPSIA) and the Centre for Security, Intelligence and Defence Studies at Carleton University for funding and hosting the workshop. My colleagues at NPSIA have been a tremendous support as I overhauled my dissertation and turned it into this book. Many others have read sections of the book and helped me refine it along the way. For this, I would like to thank Marion Barthelemy, Paula Caballero, Pamela Chasek, Felix Dodds, Judith Goldstein, Peter Haas, Thomas Homer-Dixon, Robert

Keohane, Walter Mattli, Andrew Moravcsik, Dane Rowlands, Catherine Stewart, and Michael Zürn. At Oxford, I benefited from a wonderful and supportive cohort. I would especially like to thank David Blagden, Nick Chan, Nicole De Silva, Francesca Giovannini, Nina Hall, Kai Hebel, Amy King, Sean Richmond, Landry Subianto, Henning Tamm, Diarmuid Torney, and Michael Urban. I would also like to recognize the excellent research assistantship of Karim Abuawad, Nathalie Chaar, Kaia Counter, Pablo Mhanna-Sandoval, Serena Rourke, and Selena Zabian. I would like to thank Guy Trangoš for his brilliant cover design, depicting our warming planet as well as the transnational connections – so vital to addressing global environmental issues – spanning it. The initial concept came from my brilliant and far more visual wife, Jee.

My biggest debt is to Jee, and my children: Theodore and Sophia. This book is dedicated to them. Without their patience, good humor, and love, this book would not have been possible. Jee experienced every bump in the road and moment of satisfaction along the way. When I was complacent, she pushed me. When I was proud, she brought me down to earth. No one has heard more about Temporal Focal Points and UN environmental politics than her. I would like to thank my parents, Christina and Bob Manulak, and brother, Daniel, for their constant love and encouragement. My mother-in-law and father-in-law, Eun Sook and Dong Chul Lim, have lovingly cared for our family as I wrote and then revised the book for publication. They housed, fed, and took care of the kids so that this book could be completed. They are generous people, and their *sa-rang* continues to amaze me. I thank God for the many blessings in my life, including my beautiful family, and for guiding and sustaining me along the way.

I would like to acknowledge the Social Sciences and Humanities Research Council of Canada for a doctoral fellowship, which made my studies possible. I would also like to thank the William Lyon Mackenzie King Scholarship for its support of my research. The London Goodenough Association of Canada funded a fantastic year of personal and scholarly growth in central London, enabling my archival research at the UK National Archives. My study also benefited from the research travel support of the University of Oxford's Department of Politics and International Relations' Cyril Foster Fund, the St. Antony's College Carr and Stahl Fund, and the Colin Matthew Fund for historical travel at St. Hugh's College. I would like to thank André Aranha Corrêa do Lago, Luiz de Andrade Filho, and the Brazilian foreign ministry for making Brazilian preparatory documents for the Stockholm conference available for this project. I would also like to thank John Haslam, Tobias Ginsberg, Robert Judkins, and the entire team at Cambridge University Press for their support during the publication process.

Abbreviations

CSD	Commission on Sustainable Development
ECOSOC	United Nations Economic and Social Council
FAO	Food and Agriculture Organization of the United Nations
G77	Group of 77
GEF	Global Environment Facility
HLPF	High-Level Political Forum on Sustainable Development
OECD	Organisation for Economic Co-operation and Development
SDGs	Sustainable Development Goals
TFP	Temporal Focal Point
UN	United Nations
UNCED	United Nations Conference on Environment and Development
UNCHE	United Nations Conference on the Human Environment
UNCSD	United Nations Conference on Sustainable Development
UNCTAD	United Nations Conference on Trade and Development
UNEP	United Nations Environment Programme
UNESCO	United Nations Educational, Scientific and Cultural Organization
UNFCCC	United Nations Framework Convention on Climate Change
UNGA	United Nations General Assembly

UNSSC	United Nations Environment Programme Session of a Special Character
WCED	World Commission on Environment and Development
WHO	World Health Organization
WSSD	World Summit on Sustainable Development

1 Introduction

Institutional Suboptimality in World Politics

The scale of the 1992 United Nations Conference on Environment and Development (UNCED), held in Rio de Janeiro, Brazil, was – and continues to be – exceptional. A total of 178 countries were represented, including over 100 heads of state or government. The summit was attended by tens of thousands of representatives from nongovernmental organizations, and hordes of media organizations covered the event. There was a dynamism and energy to the meeting that continues to resonate. Following the Rio conference, governments adopted a set of recommendations that targeted the implementation of an ambitious action agenda, including the creation of a high-level Commission on Sustainable Development and the reinforcement of the Global Environment Facility as a financial mechanism to further implementation. Three multilateral environmental conventions would emerge from the Rio process.

For international relations theorists, the timing of the dramatic institutional change seen in 1992 is puzzling. Many of the cooperative challenges tackled at the Earth Summit were not new. Most had been evolving progressively for at least a decade. There was no exogenous shock, crisis, or crossing of an ecological threshold that persuasively explains change at UNCED. Organizational limitations in UN environmental institutions had been long acknowledged by governments. The governance solutions adopted, which amounted to a series of steps to institutionalize sustainable development within the UN system, had been prominent for at least five years. There was no major shift in consensual knowledge or environmental ideas immediately prior to the conference that explains the sudden embrace of sustainable development. Indeed, the chief rationale for the timing of the conference was the twentieth anniversary of the 1972 United Nations Conference on the Human Environment (UNCHE, or the "Stockholm conference"), held in Stockholm, Sweden. Yet, that landmark event proved an irresistible point of focus for governments in addressing long-standing limitations in UN environmental institutions.

At least from the perspective of the global problem structure, there was therefore an arbitrariness to institutional change at UNCED. The timing of change was explained principally with reference to a conspicuous temporal landmark – the anniversary of the Stockholm conference – rather than the availability of incentives to realize change. Far from an aberration, this pattern has been common within the international system. Further governance changes in UN environmental affairs would be adopted twenty years later at the 2012 United Nations Conference on Sustainable Development, again held in Rio de Janeiro, Brazil. In case the temporal significance of 2012 was overlooked, the conference was widely dubbed "Rio+20," underlining the important anniversary. Preparations are now underway for a commemorative "Stockholm + 50" conference, to be held in the Swedish capital in 2022. This pattern extends well beyond the value of anniversaries as vehicles for coordination. "Focusing events," "demonstration effects," and prominent dates, such as a new millennium, can enable action on persistent cooperation problems in global affairs.

The Rio experience holds lessons for the current international landscape. Among the biggest challenges facing the international community is the need to update the array of international institutions that have upheld world order since the end of the Second World War. As states grapple with shifting power dynamics and a pressing set of transnational challenges, the necessity of modernizing the global cooperative infrastructure has only grown in urgency. Yet, many international institutions have proven highly resistant to change. An inability to respond swiftly to altered global realities has produced widespread institutional suboptimality. At the time of writing, ongoing efforts to increase coherence in UN environmental affairs have stretched on for decades. With the proliferation of multilateral environmental agreements, fragmentation has increased and has fueled calls for enhancing the coordinative capacity of the UN Environment Programme (UNEP). Gaps in international environmental law have been widely noted, yet have proven challenging to address. Outside of the environmental sphere, efforts to reform the United Nations Security Council lack impetus. Talks to update World Trade Organization rules to address the increased participation of state-owned enterprises in trade or the implications of digital trade barriers struggle to gain momentum. Persistent suboptimality makes these institutions easy targets for criticism from political actors seeking to curtail their country's participation in international fora.

As Stewart M. Patrick has argued, world order "is starting to crack."[1] These cracks have been widened in the largely uncoordinated international response to the novel coronavirus pandemic in 2020. Great power competition and strategic posturing, particularly between the United States and China, badly hampered the response of the World Health Organization, the G-20, and the UN Security Council. Calls for reform of the world's cooperative architecture have grown louder as a result, with many wondering whether the combined global economic and public health crisis might serve as a turning point for realizing large-scale institutional change.[2] No such change has been forthcoming. While some institutions have made incremental adaptations to the changing international context, including through layering and informal adjustments, many have faced difficulties in realizing larger, non-incremental change. The bigger reforms that can renew institutional arrangements and bring them into better alignment with changed global conditions have proven difficult to achieve, resulting in persistent governance gaps, inefficiency, and dysfunction.

Those seeking to explain big institutional transformations have focused predominantly on exogenous shocks, such as wars and depressions, as the principal drivers of institutional change.[3] Yet, these instances are – thankfully – rare. More typically, the international environment evolves gradually, with even large-scale changes occurring over a number of years.[4] Global climate change has, for example,

[1] Patrick 2018. [2] Boughton 2020; Kinsman 2020; MacMillan 2020.

[3] Interest in the subject of institutional change within the international relations (IR) field has grown rapidly since Orfeo Fioretos' 2011 article in *International Organization*, "Historical Institutionalism and International Relations." In it, Fioretos shows that, while few IR scholars had explicitly embraced historical institutionalism (HI), an approach prevalent in comparative politics, a number of analysts have conducted research that is consistent with the HI tradition. Fioretos' article sets the agenda for growing focus on the subject, and a number of more sustained treatments have recently emerged. See Fioretos 2011, 367–399.

[4] Even the rapid rise of China and other emerging powers, with important implications for international cooperation, has unfolded over decades. Big military shifts with huge ramifications for global strategic stability and arms control, such as the introduction of hypersonic weapons systems, influence global relations in a progressive fashion. Indeed, events commonly referred to as shocks may be the product of longer-term processes. Therefore, when changes emerge

been long regarded as a "creeping crisis" that has lacked a dramatic event that might break the coordinative logjam hindering more decisive international action.[5] In the North American context, progressive accumulations of sulfur dioxide and nitrogen oxides in the atmosphere contributed to transboundary acid precipitation. Yet, the Canadian government struggled for decades to generate a political focus in the US government to tackle this progressively worsening problem.[6]

The current conundrum facing many international institutions is not an absence of exogenous shocks. It is an inability to adapt to important, yet progressive, shifts in global conditions. I argue in this book that this more typical scenario is a special challenge, because it presents important coordination dilemmas for international actors seeking to determine *when* to engage in collective efforts to alter institutional norms, rules, and procedures. Such efforts, furthermore, involve a wider spectrum of global actors than is often assumed, exacerbating coordination challenges. By theorizing these coordination dilemmas and developing a fuller conceptualization of the role of temporal factors in world politics, this book explains how international actors are able to concert their expectations to address gradual, yet highly significant, shifts in the international context. Since shifts are a constant feature of international life, and because an inability to act in the face of accumulating changes is precisely the situation that global actors encounter today, such a theory is highly relevant to current policy debates.

My analysis emphasizes the role of convergent expectations in time for the successful resolution of coordination dilemmas. The availability of incentives to alter institutions is an insufficient explanation for change. Instead, actors must concert their behavioral expectations on a discrete moment in time, thereby increasing significantly their political and analytical investments in institutional change processes. These investments alter significantly the informational and bargaining conditions within an institutional setting, allowing actors to clarify the

more gradually, as is typical, current theory is poorly situated to explain when and how institutions are eventually brought into line with altered global conditions.

5 Young 1989, 371–372.

6 For a detailed account, covering this day-to-day effort from the former Canadian ambassador in Washington, see Gotlieb 2006.

nature and intensity of their preferences. In such contexts, the suboptimality of the institutional status quo relative to available institutional alternatives comes into sharper focus, increasing the likelihood of change. In addition to exogenous shocks, which provide a ready vehicle for coordination, actors often use prominent temporal landmarks, such as anniversaries and focusing events, as a means of advancing coordination. I term these moments in time Temporal Focal Points, a concept that I detail in Chapter 2.

The Puzzle

To better understand how actors respond to shifts in global conditions, we need to first examine patterns of large-scale institutional change. Despite observed difficulties, actors *are* at certain points in time able to overcome the path-dependent quality of institutions and coordinate substantial changes to cooperative structures.[7] What allows them to do so? As Giovanni Capoccia and R. Daniel Kelemen note, the emergence of critical junctures, brief phases of flux where structural constraints on institutional actors are relaxed, can open the door to more dramatic institutional change.[8] Yet more research is required on how, when, and why these junctures arise and what factors hinder their occurrence in global politics. This process has begun with Hillel David Soifer's research on productive and permissive conditions for critical junctures.[9] Future scholarship building off of this work might shed further light on questions of timing and causal underpinnings. Similarly, recent studies by Eric Helleiner and Kathryn Sikkink suggest that critical junctures may be of longer duration in international affairs than in domestic contexts.[10] This research is suggestive of the distinct challenges associated with coordinating change at the international level and provokes important questions concerning the fundamental enablers and impediments to more seamless institutional adaptation to shifting global conditions.

[7] These moments have been the subject of extensive analysis by political scientists and continue to fascinate. See Collier and Collier 1991; Eldredge and Gould 1972; Kingdon 1984; Welch 2005.

[8] Capoccia and Kelemen 2007.

[9] Soifer 2012, 1572–1597. Giovanni Capoccia has also revisited the topic, adding further insight into the concept in Capoccia 2016.

[10] See Helleiner 2017; Sikkink 2017.

While the arrival of critical junctures is theorized to have an unsystematic, random quality, an analysis of major institutional changes using a temporal lens reveals an important pattern. In many institutional settings, change is associated with high-visibility dates of purely symbolic importance to sustaining cooperative equilibria. Indeed, more than a third of UN environmental treaties of global scope have been concluded within two 2-year bursts of institutional hyperactivity in the early 1970s and early 1990s.[11] Important anniversaries have been used as common coordination points for major institutional changes in global environmental governance. The 1992 Earth Summit, for example, which was organized to coincide with the twentieth anniversary of the 1972 Stockholm conference, helped to generate momentum toward agreement on the Convention on Biological Diversity, the UN Framework Convention on Climate Change, the UN Convention to Combat Desertification, and a set of Forest Principles. Such changes resemble a "bursty" power law distribution.[12] The United Nations has relied heavily on conspicuous dates when advancing change in other areas as well. The Millennium Development Goals and the Sustainable Development Goals, two of the most significant efforts in recent memory to advance intergovernmental engagement and interagency coordination, were tightly associated with conspicuous dates. In the former case, it was the approaching millennium, and, in the latter, it was the UN's seventieth anniversary. These are combined with significant reform pushes in 1995 and 2005, which were the UN's fiftieth and sixtieth anniversaries, respectively.[13]

[11] Manulak 2020.

[12] On bursts and the power law distributions that govern many areas of human endeavor, see Barabási 2011. In the security realm, the twenty-fifth anniversary of the coming into force of the Nuclear Non-proliferation Treaty helped to generate the political momentum required to extend the treaty indefinitely. The tenth and fifteenth anniversaries of the Proliferation Security Initiative were similarly used to update the cooperative agenda and modernize the initiative (Manulak 2021). This matches the findings of Baumgartner et al. (2009) in the policy realm, who note a lack of proportion between social change and policy responses.

[13] International history is replete with such examples. While a more systematic measure of all institutional changes would be desirable, no such comprehensive list of changes exists. Indeed, what is regarded as a significant institutional change is an empirical question based in part on detailed assessment of the history of particular institutions. While the examples described in this paragraph are prominent ones, they are by no means exhaustive. I list them for illustrative

Relatedly, numerous analyses have noted the catalytic role of crises in institutional development. Indeed, in international relations, wars, depressions, and major crises have arguably been the most important drivers of large, discontinuous change. The formative role of big, rare, punctuated equilibria in patterns of institutional stability and change has been long recognized.[14] Punctuated equilibrium theorists, such as Bryan D. Jones and Frank R. Baumgartner, have examined the often abrupt nature of policy change in the wake of crises.[15] Perhaps most prominently, G. John Ikenberry has assessed the impact of major wars in allowing victors to establish new institutional orders and organizations that can persist for decades until the next great-power war triggers a breakdown of the status quo and provides an opening for constructing a new global order.[16] Exogenous shocks cause changes in critical variables that effectively clear the deck of old, self-reinforcing arrangements, thus enabling significant change.

Though some crises are produced by dramatic transformations in international conditions, known as exogenous shocks, crises frequently derive their power from the coordinative impetus that they generate. "Demonstration effects," for instance, highlight the consequences of preexisting deficiencies in global regulatory structures.[17] Often, the weaknesses uncovered by demonstration effects are long-standing, yet they trigger widespread analysis and scrutiny of existing arrangements. Analysts in the policy realm have drawn attention to the role of "focusing events" in stimulating a search for alternative arrangements and opening "windows" of opportunity for substantial alterations to the status quo.[18] The discovery of a "hole" in the ozone layer in 1985, for example, brought to light the consequences of progressive reliance on chlorofluorocarbons for the earth's stratosphere. This focusing event helped generate widespread public interest and heightened support for further research on the problem.[19] While the environmental sphere may be susceptible to this pattern, in part because public understanding of many environmental problems is mediated by the accumulation of scientific knowledge, other spheres of institutional

purposes only to demonstrate that there is a nonrandom factor associated with large-scale institutional changes over time.

[14] Krasner 1984, 242. [15] Jones and Baumgartner 2005.

[16] Ikenberry 2000. [17] Mattli and Woods 2009, 25–26.

[18] Kingdon 1984; Tomlin et al. 2007. [19] See Haas 1992a, 203.

activity are also prone to the impact of highly visible events on institutional development.[20]

The frequent association between moments of symbolic importance for recalibrating institutional equilibria, such as anniversaries and visible crises, and periods of significant institutional change holds clues for understanding phases of institutional flux and dynamism. Such deviations from the random pattern anticipated by existing theory are suggestive of underlying drivers of human behavior that demand further exploration. Common to conspicuous dates and visible moments of crisis is a widespread focus on the status of institutional arrangements. There is an associated openness and, indeed, a demand for institutional analyses and alternatives. These brief phases alter the inertial course of institutional life, leading a wide spectrum of international actors to make substantial political and analytical investments in institutional affairs.

This pattern is not easily explained by existing institutional theory. Rational choice institutionalists maintain that institutions track efficiency. State preferences are exogenously determined and respond to shifting opportunity structures. When international conditions change, institutional suboptimality can result. This can create incentives to revise institutional norms, rules, and procedures, thus enhancing efficiency. In a world of full rationality and complete information, deciding on a new institutional equilibrium in this context, while complex, is seamless. Existing institutional structures exercise little influence on the selection of alternative equilibria. Path dependence does little, for instance, influence institutional choice. Lingering at suboptimal equilibria is irrational, akin to adopting Pareto-inferior agreement alternatives. For rational choice theorists, therefore, change is generated

[20] While I have focused on environmental examples, this pattern extends beyond the environmental realm. The 2008 global financial crisis, for example, highlighted many long-standing structural deficiencies associated with the regulation of the derivatives market (Carney 2013). These deficiencies were not new, but the crisis shone a light on an existing problem. In the North Atlantic security realm, the murder of former Czechoslovak Foreign Minister Jan Masaryk in 1948 by communist forces had an "electrical" effect in spurring Atlantic countries to strengthen security cooperation and, ultimately, to produce the North Atlantic Treaty (Jebb 1972, 213). Masaryk's death was important because it came to symbolize the wider threat to European security. It did not, however, alter fundamentally the global conditions that motivated the North Atlantic Treaty negotiations.

chiefly by exogenous shocks, such as a sudden transition in the balance of power, major technological change, or the emergence of new information about cooperation problems. The *when* of institutional change is secondary, therefore, to the *why* of institutional change.

A focus on the role of nonmaterial factors, such as norms and ideas, in producing change raises still further questions related to timing. Sociological institutionalists in the international relations field, such as Martha Finnemore and Kathryn Sikkink, place analytical emphasis on the activities of norm entrepreneurs seeking to precipitate a "cascade" in which new norms take hold and, ultimately, are adopted within national and international institutions. Institutional change can immediately precede or follow norm cascades brought about by growing support for emergent norms.[21] Analysts in this tradition also examine the role of epistemic communities and bureaucratic "pathologies" in shaping institutional dynamics.[22] Other studies focus on changing intersubjective "logics of appropriateness" as critical manifestations and drivers of normative change.[23] Yet these accounts provide little insight into when one can expect norm entrepreneurs to break through, when new principled beliefs might affect institutions, or when one might anticipate abrupt change to intersubjective understandings.[24]

Historical institutionalists have given greater emphasis to the question of time. The dominant stream of this literature has conceived of timing and temporality in terms of sequence. Events that happen earlier in a causal chain can have long-term effects, locking in path-dependent processes that, once established, make institutions difficult to alter. Even when institutions change, preexisting structures affect selection among new modes of cooperation. It is within the historical institutionalist tradition that literature on critical junctures has grown and insights on more gradual modes of institutional change have emerged.[25] Despite considerable conceptual innovation, these accounts do not provide a clear means of explaining the temporal patterns in institutional change detailed in this chapter. Yet such patterns are suggestive of temporal dimensions in institutional affairs that are well in line with the historical institutionalist tradition.

21 Finnemore and Sikkink 1998.

22 Barnett and Finnemore 1999; Haas 1992b.

23 March and Olsen 1989.

24 For a comparative discussion of rationalistic and sociological approaches, see: Keohane 1988.

25 Mahoney and Thelen 2010.

Rational choice, sociological institutionalism, and historical institutionalism thus help to illuminate key dimensions of the puzzle facing institutional theorists. For received theory, factors triggering change in institutions, whether they are exogenous shocks, critical junctures, or a transformation of the normative order, should be distributed randomly across the temporal continuum. Entropy should reign. Existing theory provides little reason to anticipate the association between conspicuous moments in time, such as anniversaries or other prominent dates, and phases of institutional flux. The clustering of agreements may be more puzzling still.[26] Observed regularities in the timing of change provide an opening to supplement existing contributions through theory development that offers a fuller conceptualization of the role of temporal factors in institutional life. As we shall see throughout this book, such a conceptualization will bring to light key complementarities in existing scholarship, thereby building inter-theory crosswalks.

This analysis has implications for both international theory and practice. For theory, the inability to account for the when of institutional change suggests a gap in theoretical understanding. Indeed, the timing of a change in a variable under study is a central observable feature in virtually any exercise in hypothesis testing, whether it is in the social or natural sciences. The observed pattern in the timing of change discussed in this chapter stimulates questions concerning the processes through which gradually accumulating social and material incentives to alter cooperative structures influence institutional development. Might some other factor mediate, for example, between altered conditions and the ability to reform institutions, explaining persistent institutional suboptimality? Do these studies suffer from omitted variable bias? If so, does this help account for the current phase of institutional suboptimality in the face of significant, albeit progressively occurring, global change? What is it about highly visible junctures that influence states? For policymakers, a focus solely on potential exogenous shocks offers little prescriptive value. How will we know that key cooperative parameters have changed, except in the presence of institutional change? Similarly, do principled ideas "float freely," or do other factors favor changes in the normative order?[27] What role do norm and policy entrepreneurs play in these moments?

[26] Manulak 2020. [27] Risse-Kappen 1994, 185–214.

To Negotiate Another Day: Coordination Problems and Institutional Stasis

Before theorizing institutional change, it is necessary first to explain the coordination problem facing states. Even a cursory look at the dysfunctional status of many international institutions suggests that states face difficulties in revising institutional structures in a timely manner. States cannot summon up the impetus in bargaining to adapt institutions to shifting international conditions, thus addressing increasing institutional suboptimality. Even when the international environment has changed substantially, opening up growing material or ideational incentives for reform, prospective gains from change prove difficult to capture. This contributes to long stretches of institutional stasis and suboptimality. Governance gaps result, and frustration with the global cooperative order mounts. This is the current conundrum facing states internationally and has been a long-standing feature of the international order.

What explains the inability of institutions to track efficiency, as anticipated by pure rational choice accounts? The answer rests in the challenges associated with coordinating institutional change at the international level. While international relations theorists often boil world order down to the maneuverings of one or two great powers, this simplification obscures as much as it illuminates. As Richard E. Neustadt has noted, "reality is not bilateral."[28] Even for relatively small institutions, where coordination problems are more manageable, complex coalitional politics almost invariably enter into collective decision processes. In large UN institutions, the challenge is far greater. The task is summarized by Christiana Figueres and Tom Rivett-Carnac, two architects of the 2015 Paris climate agreement: "Bringing about a complex, large-scale transformation is akin to weaving a tapestry of elaborate design with thousands of people who have never woven anything or even seen the pattern."[29] The difficulty associated with building support for reform efforts is exacerbated further by the status quo orientation of institutions themselves, the nature of the material and ideational incentives motivating change, and the high transaction costs of these efforts.

[28] Neustadt 1970, 5. [29] Figueres and Rivett-Carnac 2020, 50.

At their core, institutions solve cooperation problems by creating a stable set of rules, norms, and procedures to tackle issues of common interest. Crucially, they extend the "shadow of the future," enabling credible commitments.[30] In doing so, institutions allow states to forego short-term benefits associated with defecting from agreed rules in favor of their longer-term interests in stable cooperation.[31] In many cases, international organizations are created to implement the resulting agreements, facilitating enforcement and providing a centralized structure to enable ongoing cooperation and decision-making.[32] These organizations themselves can become a source of institutional inertia, working to preserve vested interests in the status quo.[33] By their very nature, therefore, institutions are about long-term stability. This quality means that bargaining can almost always be pushed down the road. The ability of institutions, even highly dysfunctional ones, to live – or, for their members, negotiate – another day can make it difficult to generate the necessary coordinative impetus in bargaining.[34]

In addition to the coordination challenges inherent in the very DNA of institutions, the nature of the incentives that states experience can further complicate coordination. Changes in external conditions come in two basic forms: exogenous shocks and exogenous shifts.[35] Shocks, well-known in the literature, involve sudden, discontinuous changes in parameters underpinning cooperation. In one swift transformation, institutions are knocked out of equilibrium, thus creating incentives for states to defect from agreed arrangements. Economic collapse, major technological change, or war are common examples of shocks. Because of their suddenness and important ramifications, shocks tend to precipitate crises. As anticipated by rational choice accounts, the coordination of institutional change is common in such circumstances. Indeed, significant institutional change often serves as the proof that a shock is indeed a shock.

Exogenous shifts, on the other hand, are more gradual in nature. In addition to sudden changes, the international environment evolves more incrementally. Like shocks, shifts undermine institutional

30 Axelrod 1981; Oye 1985.

31 For the foundational theoretical text, see: Keohane 1984. See also: Keohane 1982.

32 Abbott and Snidal 1998.

33 Barnett and Finnemore 1999.

34 Indeed, as James D. Fearon (1998) observes, the long shadow of the future associated with institutional bargaining can make change even more difficult to achieve.

35 Grief and Laitin 2004; Manulak 2020.

equilibria and can provide incentives for states to alter cooperative structures. A new technology, for example, may have important implications for the health and stability of natural systems, yet these implications may not emerge all at once. Instead, certain aspects of the technology may mature over time, its applications may not be fully utilized at first, or actors may only adopt it progressively. Although they have been largely overlooked in the literature, the impacts of exogenous shifts on the preservation (and change) of cooperative equilibria are every bit as significant as those of exogenous shocks. An important difference between the two, however, is that more gradual sources of change lack an obvious coordination point in time. This complicates the coordination problem facing states in revising institutions.

Gradual shifts in the international context are a continuous feature of institutional life. The international environment has changed considerably in recent decades, yet the number of identifiable shocks is small. Indeed, while punctuated by occasional dramatic change, global social, economic, and environmental conditions are in constant transformation. The international balance of power does not usually change overnight, for example. Instead, long-term economic trends can fuel military investments that, over a number of years, produce the relative military advantages that challenge the preeminence of leading states and the global order maintained by them. With no obvious coordination point in time, the incentives to take these cumulative changes into account in institutional designs can remain unexploited for long stretches.

Growing incentives to alter institutions are also generated by self-undermining processes or reactive sequences that progressively narrow the range of parameters in which an institutional equilibrium can be sustained. Self-undermining processes take hold when factors endogenous to the institution weaken it and render it more susceptible to change.[36] Similarly, reactive sequences are, according to James Mahoney, "chains of temporally ordered and causally connected events" that "are marked by backlash processes that transform and perhaps reverse early events."[37] The logic of reactive sequences has been developed further by Tine Hanrieder and Michael Zürn, who examine the impact of power-outcome decouplings and growing

[36] Grief and Laitin 2004. [37] Mahoney 2000, 526.

authority-legitimation mismatches. Whether it be through a poor fit between powerful actors and rules or perceived legitimacy deficits, such sequences lead the losers of the institutional status quo to challenge existing rules.[38] Such a perspective is complemented by the work of Phillip Y. Lipscy, who argues that patterns of institutional change depend on the underlying characteristics of the policy area that an institution occupies. Competitive realms of institutional activity are more prone to distributive change.[39] Yet, the severity of distributive competition and the viability of outside institutional options often evolve progressively. What is critical to note about all of these processes, however, is that, like exogenous shifts, they tend to unfold over time. They do not possess obvious or ready-made temporal landmarks that ease coordination problems.

The long-term ethos of institutions and the gradual quality of most changes to international conditions make it difficult for international actors to coordinate their expectations in institutional bargaining at discrete moments in time. This challenge is exacerbated by the steep political and analytical investments required to realize institutional change. Institutional bargaining involves a significant increase in transaction costs. States must assess the cooperative environment, analyze the negotiating position of other players, and devise a new set of feasible norms, rules, and procedures to guide future cooperation.[40] Revised institutional structures can bring unintended consequences, a danger that ups the analytical demands and risks associated with change processes. Global talks can be protracted, involving years of inconclusive bargaining that demands significant resource investments in negotiating efforts. While institutional bargaining is underway, many of the benefits of ongoing cooperation can be disrupted. The energy and attention of international secretariats must, for example, be redirected to support intergovernmental negotiations. Key institutional initiatives may be shelved. Failed negotiations can create new areas of contestation and division among member states, thus reducing the credibility of long-standing commitments. Discord can grow in a manner that would be detrimental to cooperation. Proponents and opponents of change, as well as the institution itself, can suffer reputational damage. Failed negotiations can also engender "deal fatigue" where bargaining failure can poison the well for future efforts to revise

[38] Hanrieder and Zürn 2017. [39] Lipscy 2017.
[40] Keohane 1982. For a discussion, see also: Keohane and Nye 1989, chapter 3.

institutional arrangements. Wary of the risks and costs of engaging in negotiations, actors may be reluctant to recommence institutional bargaining for the foreseeable future, even when there are incentives to do so.

Heightened intergovernmental transaction costs are only part of the price of institutional bargaining. To fully assess the implications of such a spell, we must "take preferences seriously."[41] Until consciously established, preferences have a latent quality. They can be understood as the preferences that an actor *would have* if pushed to articulate them.[42] Yet, governments often have strong incentives *not* to clarify their preferences, since this can provoke domestic fights. The process of specifying and clarifying state preferences in intergovernmental negotiations is, if anything, more resource intensive at the national level than at the international level. Typically, government negotiators must obtain a bargaining mandate sanctioned by cabinet – establishing negotiating positions, strategies, and objectives. In most countries, space on the cabinet agenda is among the scarcest of government resources. Such cabinet-level processes involve intensive interdepartmental and interagency debate and deliberation.[43] These are often bruising affairs involving internal compromise and negotiation that forces governments to weigh competing priorities associated with multidimensional questions of foreign policy.

If successful, alterations to international agreements – as is typical with non-incremental institutional change – need to be ratified by national governments. Amendments to the United Nations charter, for instance, must be ratified by member states before they come into effect. These processes can be fraught with difficulty, involving negotiation with national legislators and domestic interests. In many instances, such engagements may resemble two-level games with simultaneous negotiations occurring at the national and international levels.[44] Beyond ratification, international negotiations may require

[41] Moravcsik 1997. See also: Moravcsik 2008.
[42] I thank Andrew Moravcsik for making this point, pushing me to more carefully think through this aspect of my argument.
[43] These conclusions align with the findings of Inger Weibust (2009) concerning the importance of centralization at the national level for the stringency of environmental policies. The more decentralized the process, the more challenging it is to adjust policies to changing circumstances.
[44] Putnam 1988.

the passage of national legislation that enacts new agreements. The drafting of legislation can involve multiple cabinet-level engagements to, at a minimum, issue drafting instructions to legal officers and ultimately to decide to support the proposed legislation. Negotiations with other levels of government (e.g. provincial, state, or municipal governments) may be necessary to shore up political backing and ensure effective implementation of intergovernmental agreements.

National-level mechanics of institutional change are complicated further by the need to gather information to provide a basis for government preferences in global talks. Institutional diagnostics that assess cumulative shifts in the international landscape and the various feasible institutional alternatives must be undertaken. This may involve detailed studies, as well as the convening of interdepartmental discussions or task forces. New bureaucratic units may be established to support the preference-clarifying process. Governmental and nongovernmental experts are consulted and their inputs coordinated. This enterprise can be very resource intensive. In national bureaucracies, where resources are always scarce, there are opportunity costs associated with such investments in preparations for institutional bargaining. Government priorities cannot be furthered, and other bureaus may be starved of resources. The assets emerging from these processes are highly specific, furthermore, to the bargaining setting and cannot be easily redeployed to serve other purposes.

Many institutional settings are enriched considerably by engagements with non-state actors, such as nongovernmental organizations, universities, think tanks, private industry, and the media. Think tanks and academic researchers can add information and analysis to international negotiations and to national preference-clarifying efforts. Epistemic communities and transnational advocacy networks contribute principled ideas to public discourses and deliberations.[45] Media organizations can bring to light key dimensions of institutional problems. Even if they do not have a seat at national or international negotiating tables, the simultaneous engagement of different types of non-state actors can serve as a considerable information and political force multiplier that affects institutional bargaining. All of these

[45] Haas 1992; Keck and Sikkink 1998.

organizations operate in environments of scarce resources, however. Choices must be made concerning which international issues to prioritize in their research and advocacy activities.[46] Coordination problems can lead potential contributors outside of government to focus on other issues and activities, depriving the institutional context of valuable information and analysis.

The engagement of governmental and nongovernmental players in institutional bargaining processes can therefore entail very substantial transaction costs. These costs are real and significant. This section has shown that the transaction costs associated with negotiating at the international level are just the tip of the iceberg. The national-level engagements necessary to support change in international institutions are often greater than those required at the intergovernmental level. Domestic bargaining, an essential step in informing any intergovernmental talks, is every bit as divisive and contentious as that seen at the global level. In assuming such costs, actors make choices that affect their ability to achieve other objectives. The approximation of states as unitary actors, therefore, has very real analytical drawbacks for understanding the coordination of institutional change. Consequently, the decision of global actors to engage in the purposeful assessment, consideration, and negotiation of institutional alternatives is not taken lightly.

Importantly, the significant transaction cost investments needed to produce change are unlikely to bring any significant returns unless a large subsection of relevant actors engage simultaneously in a parallel effort. Making such investments when others are not similarly invested is unlikely to bring substantial returns and entails sizable costs. It is this common enterprise that makes the large political and analytical investments involved in institutional bargaining worthwhile. There is significant interdependence at play here. A large proportion of relevant actors must engage, or none should.[47] Put another way, the steep costs involved in institutional bargaining efforts mean that actors are not incentivized to invest in efforts to alter institutions *until* they expect others to do so. The costs are too high and the potential payoffs too uncertain.

[46] Carpenter 2011.

[47] This is analogous to Thomas C. Schelling's concept of "k-groups." See: Schelling 1973.

When Expectations Converge

State and non-state actors have to bear substantial costs in the pursuit of institutional change. In view of the wide spectrum of independent actors involved in institutional change processes, the challenge confronting states in seeking to update global institutions should be understood as an exercise in decentralized coordination. A key commonality seen in scholarly accounts of punctuated change, such as critical junctures or policy windows, is a period of intense, widespread focus by multiple actors on potential advantages of change relative to the institutional status quo. Despite challenges, actors are able to engage simultaneously during these brief phases of flux in the life of an institution, opening up a window for substantial investments in change processes by state and non-state actors alike. In all of these accounts, coordinated political and analytical investments among multiple players open the door to far-reaching, non-incremental change.

Actors are able to summon up the ability to coordinate their activities in support of change processes. Expectations among relevant actors surrounding institutional bargaining converge on a focal time frame. Although actors *can* negotiate another day, they behave *as if* major temporal discontinuity looms ahead. Working backward from that timeline, they make major investments in change processes. States examine institutional options, evaluate critically the cooperative status quo, and gather information on the bargaining aims of others. Consultations with key stakeholders – national and international – are arranged. Governments solidify and clarify their preferences. Interest groups and latent national-level coalitions overcome coordination problems to mobilize as a means of influencing domestic and international deliberations. Such mobilization can influence the political priorities of governments. Norm or policy entrepreneurs marshal their political, persuasive, and informational resources to maximize their impact at such junctures.

These types of coordinated activities, occurring within a discrete time frame, amount to a sudden willingness from actors to bear significantly heightened transaction costs. For rational choice theorists, one of the principal rationales for creating institutions is the long-term reduction of these costs. Yet, the willingness to temporarily experience increased transaction costs is a precondition for institutional change. Transaction costs are, in many respects, the *lifeblood* of institutions.

While institutions exist to reduce transaction costs, the willingness of states to tolerate short-term increases in these costs is a vital factor in ensuring an institution's longer-term maintenance and viability. To alter institutions, transaction costs must be accepted.

Convergent expectations have at least two implications for institutional change processes. First, coordination brings a major injection of information, analysis, and ideas into the institutional setting. Diagnostics are undertaken and institutional alternatives evaluated.[48] The number of non-state actor inputs grow, thus contributing new information and analysis to institutional questions. This jump in the quality and extent of information available helps to clarify the state of the world. New problems and cooperative possibilities come into view, allowing for radical shifts in bargaining. Information gathering, analysis, and – more importantly – the weighing of different priorities allows states to clarify their preferences. The specification and clarification of preferences is an essential step to purposeful institutional bargaining. Diplomatically, the absence of fully articulated preferences, typically manifested most concretely in a set of instructions from capital, results in a defensive and conservative approach to bargaining that reinforces the status quo. Without the expectation of widespread coordination, these types of investments are difficult to justify in perpetually resource-starved government contexts.

Coordination and clarification of national preferences allows governments to update their information on the bargaining positions of others. Throughout negotiations states possess probabilistic estimations of what is possible at the bargaining table, based on the quality of information that is available about the negotiating stances of other players. The better the information available, the narrower the range of possible outcomes that can be entertained. Conversely, when the attitudes of key states have not been examined recently and national preferences are unclear, states must consider a much broader array of possible bargaining outcomes.

Second, convergent expectations help to generate a significant mobilization of latent political forces, nationally and internationally. These forces include political parties, interest groups, think tanks,

[48] A similar observation is made by Jacques Fomerand and Peter M. Haas in their analysis of the impacts of United Nations conferences, see: Fomerand 1996, 366–368; Haas 2002, 84–85.

bloggers, and opinion-shapers of all stripes.[49] The widespread coordination of expectations surrounding institutional bargaining can facilitate the organization of domestic social support for updating institutional structures, particularly in democratic countries. It is often costly and difficult for social groups to identify potential allies in promoting change in international institutions. The organization of potential advocacy efforts is costlier still. Similarly, norm and policy entrepreneurs both inside and outside of government may be waiting for a propitious moment to target their social advocacy investments. A temporal focus makes it easier for political actors to devote the resources necessary to embark on coalition building and advocacy. In this way, convergent expectations can help to overcome the aggregation or representation failures that lead states to underestimate domestic levels of support for institutional change.[50]

The expectation of serious negotiations along a discrete timeline can also lead to the creation – at least on a temporary basis – of new bureaucratic structures and interests that can affect the political balance nationally and internationally. In addition to new information, therefore, the launch of task forces or new bureaucratic processes can generate new actors that have a stake in global discussions and domestic preference formation. These actors influence the preference clarification process and may serve on national delegations to international meetings. They provide assessments and briefing materials that can add considerable impetus to wider efforts to renew institutional structures. Political leaders, furthermore, will often observe negotiations from the sidelines until the likelihood of an agreement seems high. When a deal seems likely, high-level political actors may jump on the bandwagon to add political weight and, perhaps, take credit.[51]

Convergent expectations, therefore, can reshape fundamentally the information environment and the balance of political support for institutional reform. Actors are willing to bear substantially heightened transaction costs to realize change because they expect others to do so. It is this expectation that makes major political and analytical

49 See also, Fomerand 1996, 370; Haas 2002, 75.

50 For a discussion of such mobilization, see: Moravcsik 1999, 283–284.

51 Former United States National Security Advisor and Secretary of State Henry Kissinger, for instance, would not participate directly in negotiations unless he thought that there was a better than 80 percent chance of success. See: Sebenius et al. 2018, 79.

investments in bargaining processes a welfare-improving proposition. To a far greater extent than existing theory allows, this is a question of timing. It is a central contention of this book that the moments of institutional flux described by analysts of critical junctures and exogenous shocks are characterized by the type of convergent expectations described above.

Convergent Expectations at the Negotiating Table

The preceding analysis has highlighted the coordination dilemmas posed by gradually emerging change in incentives and the critical role of convergent expectations in realizing institutional change. I now show how and why the arrival of convergent expectations increases the likelihood of change in international institutions. The injection of information and analysis, as well as the mobilization of latent political forces, can reshape the negotiating setting. Bargaining theory illuminates the impact of these factors on the likelihood of institutional change. First, new information and the mobilization of political forces can influence the intensity in which states hold their preferences for particular institutional alternatives.[52] New information can uncover the extent of the incentives associated with swiftly adapting institutional structures to new realities. The mobilization of latent social forces can give actors a domestic political stake in change efforts. The strength of preferences is a vital factor in the consummation of issue linkages, a vital means of achieving joint gains in talks. To entertain linkages, actors must conduct a comparative assessment of the importance that they attach to each potential agenda item. Actors may be willing to make concessions on one issue in order to obtain them on another that they regard as more important. In this scenario, both sides leave the bargaining table better off.

A second way that the bargaining setting is impacted by new information and mobilization is the clarification of preferences. With conscious scrutiny, states may find that they have definite attitudes toward risk or strongly held forecasts related to unknown future contingencies. They may realize that they discount the future in favor of short-term gains (or the reverse). Similarly, they may value symbolic

[52] It is also conceivable that the intensity of preferences can diminish. States may conclude that they are less vitally interested in an issue than previously thought.

achievements, precedents, or ideological coherence.[53] The shape of states' preference functions cannot be discovered without analysis and transaction cost-intensive domestic deliberations that open the door to intergovernmental agreements that were not previously achievable. "Dovetailing differences" in risk attitudes, forecasts concerning unknown future contingencies, and time preferences can provide the building blocks for joint value creation and cooperative dealmaking. States may, for example, build agreements to share risk, thereby adding value for others that are risk averse. More risk acceptant states may seek concessions in other areas. In another scenario, states that value current, more than future, gains may make concessions on long-term aims in favor of more immediate benefits. If preferences remain latent, these means of expanding the payoff pie may well have gone unnoticed.

As information accumulates on the bargaining position of other actors, dealmaking opportunities come into view that were not previously seen. New information also helps states to update and sharpen their probabilistic estimations concerning the range of scenarios possible in talks. They can, in many instances, rule out certain negative outcomes, thereby effectively solving what can be termed the "can of worms problem," where states avoid reopening institutional rules and procedures to negotiation for fear of ending up worse off than they were before. While this process may also lead states to rule out some very optimistic bargaining outcomes, prospect theory suggests that states weigh potential losses more heavily than prospective gains.[54]

The third and perhaps most important implication of convergent expectations at the negotiating table is that they help to overcome bargaining failure. In talks that lack a clear and temporally distinct endgame, as is the case with current reform efforts surrounding the UN Security Council, states can be caught in the temporal equivalent of what James K. Sebenius terms the "negotiator's dilemma."[55] In the negotiator's dilemma, all sides would benefit from value-building, integrative bargaining. By sharing information on preferences and engaging in a relatively open manner, states push the Pareto-frontier outward and, as a consequence, enjoy superior agreement possibilities. If either side succumbs to the temptation to withhold critical information or to misrepresent their true preferences, however, the more

[53] Fisher et al. 1991, 74. [54] Levy 1997, 87–112. [55] Sebenius 2002, 241.

cooperative side can be exploited in negotiations. It is this risk that often leads to suboptimal outcomes that fail to maximize joint gains. This dynamic increases the probability of bargaining failure in the face of achievable gains or Pareto-inferior agreements.

A temporal divergence of expectation works in roughly the same manner. In this version of the negotiator's dilemma, states must decide *when* to engage integratively – exchanging information openly and making concessions. Absent a focal time frame, bounding the scope of talks, negotiators face difficulties in sequencing their concessions. Concessions made could be pocketed by the other side without reciprocity. The emergence of a focal time frame for talks adds coherence to this process, allowing actors to move progressively beyond posturing toward their true bottom line. This increases the probability that states will identify common ground and maximize joint gains. By specifying and solidifying their preferences, furthermore, states conduct the types of analysis necessary to create value in talks. The capacity to do so is especially important in achieving institutional change, since it is in these contexts that many states possess a veto over possible amendments. The ability to expand the payoff pie, all things being equal, makes successful negotiations much more probable. In addition, agreements that bring value to all sides are more likely to be durable.

Achieving Institutional Change

International actors face great difficulties coordinating large, nonincremental institutional change, leading to a status quo bias and suboptimality. Part of the coordination challenge is baked into institutions themselves, which incentivize states to orient their behavior in relation to a long shadow of the future. Even as global material and ideational conditions shift, opening up incentives to adapt institutional arrangements, many of these changes are progressive and provide no obvious coordination point in time to galvanize institutional bargaining.

Coordination difficulties are made more severe by the high transaction costs involved in altering institutions, both at the national and international levels. Institutions exist to reduce short-term transaction costs, but the willingness of international actors to assume them in the immediate term is vital to realizing change. Substantial, coordinated political and analytical investments by a large proportion of the institutional population provide vital information about the state of the world and illuminate essential features of the bargaining context. Convergent expectations can facilitate the mobilization of latent social forces that influence the national and international political setting. Coordination also gives structure and coherence to negotiating processes, allowing actors to overcome temporal dimensions of the negotiator's dilemma.

Convergent expectations are critical to altering institutional structures through the clarification of state preferences. By recognizing the intensity of their preferences, states can assign an appropriate value to possible institutional bargains, enabling issue linkages. As states develop a fuller appreciation of the shape of their preference function, including risk attitudes and time preferences, they are better able to dovetail differences to achieve joint gains. The capacity to expand the payoff pie is particularly crucial in bargaining contexts replete with veto points, such as those associated with altering institutional rules and procedures. In this way, the capacity to create value in negotiations enables institutional change.

By assessing the potential ramifications of temporal coordination problems and giving greater emphasis to processes of national preference clarification, this assessment sets out very different expectations than existing theoretical accounts. Rather than assuming that temporal coordination is driven mainly by the presence of incentives to realize change, the preceding account finds that coordination problems constitute a more important barrier to change than previously recognized. It suggests that actors' willingness to assume high transaction costs varies over time. The prospect of convergent expectations opens the possibility that political and analytical investments in change processes may bring returns, opening the door to institutional transformation. Thus, the achievement of coordination – a temporal convergence of expectations – is a necessary condition for punctuated change.

This does not imply that the presence of incentives to alter institutions is a secondary factor in explaining change. Quite the contrary. The presence of such incentives constitutes one of two independent variables considered in this analysis. As will be presented in detail in Chapter 2, the benefits of change rest heavily on the strength of available institutional alternatives compared to the status quo. At the same time, it suggests that, even in the presence of such incentives, states may have difficulty coordinating to realize beneficial change. Focusing attention on the question of why the timing of change does not seem to correspond closely to the emergence of change incentives, the preceding analysis finds that superior agreement possibilities are frequently available. Given the lags between exogenous shifts and institutional change, this presumption is not an unreasonable one. The accumulation of cooperative potentialities is a common feature of institutional life and is a necessary condition for temporal coordination problems to matter.

Although the preceding discussion has concentrated on material incentives for change, it complements the assessments of analysts focused on social factors in institutional development. These accounts assume that the arrival of new principled ideas or a norm cascade will typically precipitate the coordination of actor expectations necessary to produce change. I problematize that assumption. Instead, I suggest that the mobilization of intellectual, normative, and ideational resources may be predicated to a great extent on convergent expectations. The types of advocacy and intellectual investments that generate normative transfigurations are made easier when convergent expectations take hold. This is consistent with the conclusions of John W. Kingdon, who finds that policy windows serve as a catalytic force for policy entrepreneurs and give rise to greater receptivity among decision-makers to consider policy alternatives. Rather than producing coordination, therefore, the development of new norms and ideas is stimulated heavily by coordination. There is, moreover, an interplay and synergism that develops when ideas are under active debate, helping to refine their content. Widespread dialogue and consideration within public discourses contributes to normative development.

The preceding assessment also suggests that a much broader set of players impact change processes than is conventionally assumed in global politics. Existing accounts tend to focus on the role of a hegemon or hegemonic subgroup as the driving force for change. The emphasis on coordination among a broad set of players is the antithesis of the hegemonic approach. The empirical evidence outlined in this book supports this concentration, demonstrating how convergent expectations among diverse actors influence change processes. Such an approach holds special interest in a period when global power is becoming more diffuse. In what Joseph S. Nye terms, "the other global power shift" of our times, a broad spectrum of actors have moved to the international fore, transforming world politics.[56] By explaining how diverse players coordinate to realize institutional change, this book provides an enriched and policy-relevant perspective on factors affecting change today.

Structure of This Book

To conclude this chapter, I provide a summary of key findings. In this book, I examine the relationship between gradually accumulating incentives to alter institutions and actors' capacity to achieve coordination as a means of capitalizing on those incentives. Could the presence of temporal coordination dilemmas – and their resolution – explain patterns of change in global politics? Could such factors account for the puzzling patterns in the timing of change described at the beginning of this chapter?

In the ensuing chapters, I show that the emergence of a temporal convergence of expectations, often triggered by the arrival of a conspicuous, unique, definite time frame, helps to explain patterns of continuity and change in international institutions. I term these moments Temporal Focal Points. I test the theoretical framework through a detailed case study of the record of continuity and change in United Nations environmental institutions. As we will see, the institutional problems caused by gradual exogenous shifts that progressively created incentives to revise environmental institutions took a puzzlingly long time to address. The optimal time for agreement on big institutional changes would appear to be earlier than occurred. The power of the three focal junctures that saw significant change in UN environmental institutions – 1972, 1992, and 2012 – was derived from the fact that they were seen as highly conspicuous, catching the eye of all relevant actors. These highly visible phases eased coordination and facilitated change.

In Chapter 2, I develop and formalize the arguments outlined in this introductory chapter. I begin by showing how the situation facing global actors resembles a stag hunt game, where players can realize beneficial institutional change if they are able to coordinate their political and analytical investments in institutional change efforts.

[56] Nye 2020.

Since the number of relevant actors in international institutions is so large – spanning hundreds of distinct domestic and international, governmental and nongovernmental, organizations – coordination is an undertaking of enormous complexity. Achieving a genuine coordination in behavior is not as simple as adding a subject to the intergovernmental agenda or scheduling a meeting. Actors must invest scarce resources in institutional diagnostics and opt to communicate their underlying interests to other parties in a relatively open and transparent fashion. If players act cooperatively when others do not, they risk ending up significantly worse off than they would have been had they not done so.

The risk-dominant nature of status quo institutional arrangements leads states to linger at inferior, payoff-dominated equilibria longer than would be predicted by pure versions of rational choice institutionalism. Since collective action on festering institutional deficiencies is predicated on a coordination of expectations at discrete points in time, actors face a classic assurance problem. Although each could achieve a better outcome if they could rely on others to act in concert with them, they face important risks in this endeavor. It is the risk-dominant character of this situation that makes change problematic, contributing to a status quo bias in institutional life. The key is reaching a temporal convergence of expectations: a point in time when all actors believe that others will behave in a certain manner. The emergence of conspicuous, unique, definite temporal signposts eases the process of coordination by providing a prominent phase along the temporal continuum where this belief can take hold. After detailing the role of two main independent variables, I discuss the principal pathways through which large-scale institutional change occurs. The chapter then specifies observable implications of the theoretical framework presented and concludes with a discussion of research design and case selection. This sets up the empirical investigation to follow.

To probe the plausibility of this theoretical framework, I undertake a detailed empirical investigation of the record of change in United Nations environment institutions from 1963 to the present. While I devote extensive attention to the entire period, my empirical chapters center on four main phases: the 1972 UNCHE, the 1982 United Nations Environment Programme Governing Council Session of a Special Character (UNSSC, or the "Nairobi conference"), the UN General Assembly's response to the release of the 1987 report of the

World Commission on Environment and Development (WCED, or the "Brundtland report"), and the 1992 United Nations Conference on Environment and Development (UNCED, or the "Earth Summit"). In Chapter 7, I supplement these detailed historical treatments with an assessment of the post-UNCED period. The diversity of the cases over time contained within this book allows for significant variation in values "assigned" to the independent variables, aiding causal inference.

Chapter 3 analyzes the Stockholm conference and the years immediately preceding it. The early 1970s represented a period of transition in UN environmental institutions. Mounting global environmental problems in the decades prior to 1972 had led to an ad hoc accretion of institutional measures within the United Nations system. While the need to address major defects in extant institutional arrangements had existed for at least a decade, actors faced sharp coordination dilemmas. Leading developed countries feared that, if they bargained cooperatively with developing countries, they risked the rise of environmentally damaging pollution havens in the global South. Developing countries feared that cooperating with industrialized countries could lead to the rise of environmentally motivated nontariff trade barriers and green conditionality in the provision of development assistance. Although all sides had much to gain from international cooperation in the environment field, this strategic dynamic hampered greatly efforts to institute new rules and procedures that could address planetary decline.

The inertial drift in UN environmental affairs persisted until 1972, when UNCHE emerged as a highly conspicuous focal juncture, or Temporal Focal Point. Though the Stockholm conference was initially seen as a relatively low-level technical affair, it quickly gathered momentum and provided a focal time frame that enabled a temporal convergence of expectations among actors, North and South. The conspicuousness of the Stockholm moment was enhanced by a string of prominent environmental disasters that occurred in the years leading up to the conference. Growing environmental concern was crystallized by UNCHE's entrepreneurial Secretary-General, Maurice Strong. As the conference neared, actors engaged cooperatively in the search for institutional alternatives that met the changed problem-setting. The result of the conference was dramatic institutional change. The Stockholm action plan and declaration provided a number of recommendations and principles that set out the basis for global environmental governance going forward. The conference also resulted in recommendations that led to the creation of UNEP by the UN General Assembly in December 1972.

Chapter 4 focuses on the decade after the Stockholm conference. In that period, UNEP compiled a mixed record in the promotion of global environmental cooperation. As the tenth anniversary of UNCHE neared, UNEP's Executive Director, Dr. Mostafa Tolba, spearheaded a UNEP "Session of a Special Character" (UNSSC) to regain the impetus of the Stockholm conference and address institutional defects that had become apparent between 1972 and 1982. The 1982 Nairobi conference became a highly conspicuous moment in UN environmental politics. The conference brought a period of intense focus on global environmental despoliation and on the weaknesses of post-Stockholm organizational measures. There was a large proliferation of institutional diagnostics and a heightened intensity in intergovernmental negotiations. UNSSC, therefore, constituted another Temporal Focal Point in the UN environmental sphere.

Although organizational limitations motivated members of the African group – led by Kenya and Tunisia – to propose institutional change at the Nairobi conference, underlying conditions had not changed significantly between 1972 and 1982. Lead states, including the United States and United Kingdom, maintained that existing institutional measures were, on the whole, sufficient to meet the scale of global environmental problems. Though some developing countries called for an operational mandate for UNEP, other developing countries were wary of the implications of institutional change for their sovereignty. While the Kenyan conference contributed to a temporal convergence of expectations among actors, no substantive basis for institutional change existed. States lacked an incentive to revise institutional arrangements. Despite the scheduling of a multilateral conference, a coalition sufficient in size and composition could not be built to secure institutional change. Existing institutional measures continued to represent states' best response to the global environmental problem structure in 1982.

The years following the Nairobi conference saw significant shifts in international conditions, as the Third World debt crisis greatly exacerbated environmental problems in the Global South. Chapter 5 assesses the implications of these shifts for UN environmental governance.

While environmental challenges, such as poor sanitation and local pollution, were present in parts of the developing world well before the debt crisis, mounting financial troubles caused countries in Latin America and Africa to accelerate environmental exploitation as a means of servicing unpayable external debts with export earnings. The massive shift in aggregate net resource transfers from South to North through the 1980s had serious environmental consequences.

The national and international economic decision-making structures that drove planetary decline were increasingly singled out as the chief cause of mounting environmental problems. Environmental institutions that focused primarily on cleaning up the ecological messes created by decisions taken within other institutions or branches of government were unsuitable for the new problem-setting. As external conditions shifted progressively, the World Commission on Environment and Development, which was formally established in December 1983, was deliberating. Noting the changed international problem-setting, the commission, chaired by Gro Harlem Brundtland, followed an "alternative agenda" that focused heavily on the economic drivers of environmental deterioration. The "Brundtland report," titled *Our Common Future*, popularized the concept of sustainable development and, among many other things, proposed a series of far-reaching reforms that would institutionalize sustainable development within the United Nations system.

The Brundtland report was presented to the UN General Assembly in October 1987. The ensuing negotiations centered on ensuring adequate follow-up on the commission's recommendations. Some states, led by Norway, sought to set in motion a detailed bargaining process to precipitate institutional change then and there. These efforts encountered problems. While the General Assembly recognized sustainable development as "a central guiding principle of the United Nations, Governments and private institutions, organizations and enterprises," follow-up mechanisms lacked coordination. This was because countries, North and South, faced important risks in embracing sustainable development. Developed countries risked being obligated to assume an onerous financial burden in the pursuit of sustainable development, a still vague concept that could be operationalized as a developing world license to pollute and to exploit their natural resources. Developing countries feared green conditionality and unjustified intrusions on their sovereignty.

Institutional change did not therefore occur in 1987. A review of states' bargaining positions shows that all sides were broadly supportive of the Brundtland Commission recommendations. Though receptive, deep caution won the day. All sides faced important risks if they embraced the sustainable development formula at the General Assembly. Temporal coordination problems were at the root of actors' slow embrace of the WCED proposals. None wanted to limit their flexibility or compromise their bargaining position in subsequent negotiations. There was, furthermore, ambiguity concerning the timeline for institutionalizing sustainable development within the UN system. Some expected follow-up negotiations to begin immediately, while others focused instead on a notional – yet still unconfirmed – multilateral conference in 1992. A temporal divergence of expectations was the enemy of purposeful negotiations in 1987. Instead of serious bargaining, states postured, unwilling to approach talks cooperatively.

Chapter 6 begins with a brief examination of the institutional deficiencies highlighted by the Brundtland Commission. The coordination problems that prevented serious bargaining in 1987 persisted into the early 1990s, when preparations for the United Nations Conference on Environment and Development began to intensify. The "Earth Summit" in Rio de Janeiro, Brazil, quickly became a conspicuous, unique moment in time that greatly influenced state behavior. The conference took on an unparalleled prominence, generated by growing public awareness of global environmental problems, the twentieth anniversary of the halcyon days of the Stockholm conference, and the efforts of the UNCED secretariat. The Rio conference became a Temporal Focal Point in UN environmental politics.

The approaching conference provided just the stimulant needed for states to address the institutional recommendations of *Our Common Future*. A series of intensive formal and informal negotiations occurred in 1991 and early 1992 that established a consensus on significant institutional reforms. Sustainable development became the guiding principle for the recommendations of the Rio action plan, called "Agenda 21," and the principles contained in the Rio declaration. Organizational measures were also altered, launching a UN Commission on Sustainable Development and providing a new mandate for the World Bank-operated Global Environment Facility. Most of these proposals were shaped by the institutional change recommendations of the Brundtland Commission.

The realization of institutional change in 1992 is so compelling because it brings the power of coordination – and Temporal Focal Points – into sharp relief. Between 1987, when the Brundtland Commission recommendations were first considered by states, and 1992, underlying conditions had not changed significantly. Existing institutional arrangements, including UNEP, were widely viewed as incapable of arresting the progressive deterioration of the world environment. Many of the ideas that formed the basis for institutional change in 1992 were derived from the 1987 Brundtland report. The main difference between the WCED and Rio cases was that, in the latter instance, states had a clear temporal focus that allowed them to reach Pareto-improving agreements. The conference added a powerful impetus to talks. The approaching Earth Summit provided a temporal point of focus for actors in their efforts to reach agreement, easing coordination problems. All saw the Brazilian conference as the time to alter institutional arrangements. They invested the necessary analytical resources in the process and made the concessions required to finalize a settlement in time for the June 1992 gathering.

Chapter 7 brings the discussion forward from UNCED to the time of writing, examining the period running from 1993 to 2021. The style of the empirical analysis in this chapter is different in character from that seen in Chapters 3–6, focusing almost exclusively on an application of the theoretical framework at the level of comparative statics. Absent the extensive available archival documentation used to enable a detailed process-based evaluation of institutional change negotiations, I focus instead on an assessment of outcomes.[57] This approach increases the number of observations of the independent variables and heightens confidence in my theoretical propositions. The chapter first examines the period running from 1993 to 2002, culminating in an assessment of the 2002 World Summit on Sustainable Development, held in Johannesburg, South Africa. While incentives to alter institutions were present, divergent temporal expectations led to the preservation of the institutional status quo.

Institutional problems – and the incentives available to address them – persisted through the first decade of the 2000s. This situation

[57] Many archives follow a thirty-year rule for the release of diplomatic documents, meaning that the documents that could inform analysis of the post-UNCED period are simply not available for public consultation.

changed as momentum built toward a Temporal Focal Point: the 2012 United Nations Conference on Sustainable Development, held in Rio de Janeiro. Convergent expectations around "Rio+20" led to the swift adoption of a series of significant institutional changes, including a suite of measures to strengthen UNEP, discontinue the failing Commission on Sustainable Development, create the High-level Political Forum on Sustainable Development, and establish a process to articulate the Sustainable Development Goals. The chapter then discusses the period since 2012, analyzing key developments in UN environmental affairs.

The concluding chapter summarizes the key findings contained in the preceding chapters, highlighting the theoretical implications of this book. The Conclusion assesses the extent to which the theoretical framework outlined in Chapter 2 hold up through the case studies contained in Chapters 3–7. The chapter ends with a discussion of some promising theoretical avenues opened up by this framework and policy implications of my analysis. It surveys the global environmental and political context at the time of writing, underlining incentives to reform current UN environmental institutions and the presence of temporal coordination dilemmas. It concludes by assessing the potential for the arrival of a Temporal Focal Point in the current context.

Conclusion

This chapter highlights some puzzling features of the record of major changes in international institutions. States face important challenges in altering institutions, particularly when incentives to revise cooperative structures accumulate gradually. In addition to reaching agreement on a particular institutional alternative, therefore, actors must first reach a convergence of expectations to address accumulating incentives at one point in time. In addition to detailing the nature of these coordination dilemmas, this chapter summarizes the principal empirical findings of this book. In the next chapter, I provide a more detailed theoretical discussion of these issues and set out the rationale for case study selection and data collection.

2 It's About Time

Explaining Change in United Nations Environmental Institutions

Despite significant transformations in the broader international system, including the emergence of pressing cooperation problems and shifts in the global distribution of capabilities, international institutions have proven highly resistant to change. When change has occurred, it has often been incremental in nature, failing to reflect the scale of the transition in the wider global system. Within UN environmental institutions, for instance, growing fragmentation saps the capacity of states to tackle cross-cutting threats to earth systems. The poor fit between leading international institutions and current realities has eroded the legitimacy of many global bodies and affected the willingness of states to contribute political, economic, and military resources to multinational efforts. As material and ideational incentives mount progressively and cooperation problems intensify, states face a growing reason to overcome impediments to coordination and realize significant institutional change.

Chapter 1 established that large-scale change requires a coordinated effort from a wider array of international actors than is often recognized. Indeed, high complexity – and the collective action problems associated therewith – may be a key differentiating feature between domestic and international political contexts.[1] The complexity of international policymaking and negotiation renders swift collective decision-making a significant challenge.[2] In addition to a large number of states, non-state and sub-state actors are important players that can affect reform efforts. While they are rarely able to block change, these actors are a latent force in institutional life that have the capacity to enrich the bargaining context in a manner that can open the door to punctuated change. They do this, principally, through the informational and political conditions shaping state preferences.

[1] Zürn 2016, 205–206. For a discussion, see: Weibust 2009.
[2] Hampson and Hart 1999.

Such complexities make coordination difficult. Temporal coordination dilemmas can prevent international actors from enjoying the accumulated benefits associated with revising institutional structures to meet changed international conditions. Such dilemmas are defined by the ongoing difficulty of getting all actors to invest in institutional change processes on a concurrent timeline, despite their self-interest in doing so. All things being equal, the larger the number of consequential players, the more difficult it is to get all actors on the same page at the same time. These actors are, furthermore, highly diverse in their size, economic situation, regime type, regional context, linguistic profile, and historical experience. This affects coordination efforts. The matter is not as simple as adding an item to the intergovernmental agenda, scheduling a conference, and setting a deadline. Formal consideration does not necessarily entail serious or purposeful bargaining.[3] For substantive negotiations to occur, there must be a convergence of expectations among actors on some point in time. If important parties are unwilling or unable to engage in serious talks, the institutional status quo will prevail. Reaching such a point is difficult, given the progressive manner in which incentives to revise institutional arrangements typically mount.

The preceding analysis suggests that reform efforts in large-n contexts can face significant difficulties. As James D. Fearon's work on institutional bargaining demonstrates, the bargaining phase of international cooperation can resemble a war of attrition game. When agreements are enforceable and the shadow of the future is long, actors are incentivized to hold out as long as possible in bargaining for the best possible agreement, incurring costs of delay, so as to secure anticipated future gains.[4] When the stakes are high, reaching agreement is complicated. Fearon's conclusions are, moreover, derived from a two-player game and thus do not grapple with the complexities experienced in multilateral institutions. When projected onto a multilateral canvas, temporal coordination challenges, not accounted for in Fearon's treatment, come to the fore. Indeed, as Lisa L. Martin notes, the resolution of assurance games can be highly sensitive to the

[3] Indeed, routinized discussions can serve as a vehicle for "counter-diplomacy," in which ongoing conversations are used as a vehicle to "evade or frustrate political solutions or international rules," while appearing constructive. See: Barston 2019, 5.

[4] Fearon 1998.

assumption that states are behaving as rational, unitary actors.[5] When actors are non-unitary and intra-state negotiations are divisive, as is so often the case when governments bargain over institutional change, coordination can be fraught. For instance, in two-player coordination games, one side can effectively "force" the other into cooperating by signaling its intentions and pre-committing to a preferred outcome. This can be difficult when different domestic factions increase the level of uncertainty about the other side's expected behavior. Such a maneuver is especially complicated in large-n contexts both because it is less clear that a sufficiently large number of players will cooperate and because, even when feasible, effective signaling can be costly. Thus, even when the aims of institutional change are primarily positive-sum in nature, focused on capturing new available gains rather than targeting zero-sum redistribution, change can be hindered greatly by coordination dilemmas.[6]

This process is further complicated by the fact that challenging or reassessing – let alone changing – institutional structures is a costly enterprise. Even the most well-resourced states and departments face constraints. To prepare for talks, governments invest heavy analytical resources for diagnosing institutional problems and devising alternatives. Information on complex issues must be gathered and intelligence on the bargaining positions of other states accumulated. Task forces may need to be established and the attention of harried decision-makers demanded. Negotiations to change cooperative arrangements are, by definition, extraordinarily relative to the normal state of affairs. The authority delegated to policy-makers through established, routine operating procedures is exceeded, thereby increasing transaction costs.

Actors wishing to call institutional structures into question invest political resources in their efforts and may forego benefits normally accruing from the institution in question. Diplomatic capital is spent seeking to build support among pivotal players and to kick-start negotiations. This valuable and limited currency is commonly purchased through vote trading, military commitments, or development assistance. By calling in diplomatic favors or making promises of future reciprocity, reformers expend scarce assets. States spearheading calls for reform also

[5] Martin 1992, 780–783.

[6] For accounts focused on redistribution issues, see: Kruck and Zangl 2020; Lipscy 2017.

invest national prestige in their effort and failure can be embarrassing. Unsuccessful initiatives affect the political fortunes of governments and result in reputational damage. Reformers also have fewer resources to pursue other policy initiatives of national importance or to exercise influence on organizations. Agenda space or voice opportunities utilized in advancing reform cannot be employed in pursuing other priorities. Far from cheap talk, these costs are real and significant.

The process of calling current patterns of behavior into question can also erode the legitimacy of existing institutional arrangements. Even when prevailing norms, rules, and procedures remain unaltered, the contested state of institutional arrangements can invite future challenges to the regime. While these challenges may be highly visible, they can also be small and almost imperceptible. As rules lose their legitimacy, for example, a gap can emerge between rules "on paper" and their implementation "on the ground." The unsettled state of affairs may support the rise of various types of change agents that can act – intentionally or unintentionally – to alter institutional practice by pursuing their distributive aims.[7] This circumstance may be the product of self-undermining processes that progressively diminish the status of institutions.

In pursuing institutional change, actors therefore assume important risks that, if others do not act in concert with them, they will end up worse off than if they had not challenged the existing order. The best response of each is to choose the course of action that it expects others to choose: If a large proportion of relevant actors do not elect to enter into serious negotiations to alter institutional arrangements, none should.[8] Getting many diverse players onto the same page is difficult. Coordination is hampered greatly by communication problems. At the international level, bargaining includes multiple parties and issues, resulting in enormous complexity. This coordination problem is, fundamentally, a question of timing. Unless relevant actors' expectations converge on a specific time when they expect others to expect to be expected to behave in a certain manner, they cannot coordinate.[9]

[7] For a discussion of the role of different types of change agents, see: Mahoney and Thelen 2010, 22–27.

[8] This does not mean that *all* actors necessarily must be onside. It does, however, mean that actors' decision of whether to act cooperatively depends on the number of actors participating. This is similar to Schelling's concept of "k-groups." See: Schelling 1973.

[9] Phrasing taken from: Schelling 1960, 57.

Table 2.1 *Matrix of temporal coordination*

		Side 2	
		T_1	T_∞
Side 1	T_1	A, W	D, X
	T_∞	B, Z	C, Y

Here, the following relations hold:
$A > B \geq C > D$
$W > X \geq Y > Z$

Convergent expectations are thus the key to resolving the temporal coordination problems associated with institutional change. Compared to exogenous shocks, which precipitate a crisis situation, exogenous shifts, self-undermining processes, or reactive sequences often lack a comparable temporal coordination mechanism.

The matrix below illustrates the choice situation facing states and relevant international actors. In the presence of incentives to alter institutions, all sides have an opportunity to coordinate their behavioral expectations on the earliest available time frame (T_1). The temporal alternative that they face is an undefined, indeterminate point in the future (T_∞). At the existing, status quo equilibrium, therefore, sides play T_∞, as institutional bargaining is not under active consideration. However, all stand to benefit from acting in concert if they are able to coordinate. Outcomes A and W are the best for each side, since they receive their largest payoff. The danger that each player encounters is that, should they seek to coordinate at T_1 and the other side opts for the indeterminate timeline, they will end up with their worst outcome (D or Z). The less risky choice for each is to maintain the existing equilibrium. By putting off negotiations to some undefined point in the future, they avoid their worst result and receive their next best outcomes (B, C or X, Y). This scenario is classically labeled a stag hunt after Jean-Jacques Rousseau's parable (see Table 2.1).[10]

[10] In the stag hunt game, the actors choose simultaneously whether to hunt a stag or a rabbit. Each is able to hunt and catch a rabbit without the aid of the other. Although rabbits provide a small payoff, the stag – even when divided between the two hunters – offers more. The appeal of the rabbit is that the hunters' ability to capture it does not rest on the actions of the other hunter. If, however, both are able to concert their expectations and hunt a stag, each side will be better off. See: Rousseau 1984, 111.

In the presence of incentives to alter cooperative arrangements, actors therefore can enjoy ∞a superior institutional alternative immediately if they are able to coordinate. The stag hunt game assumes the existence of a uniquely efficient solution to the temporal coordination dilemma described above. This is because, in the presence of available joint gains, states receive their maximum possible payoff by moving immediately to introduce alternative institutional arrangements.[11] Lingering at payoff-dominated equilibria is, from this perspective, suboptimal. The payoffs listed in the matrix above do not, therefore, correspond to those associated with different agreement possibilities, as is conventionally the case in a game theoretic setup. Actors are not coordinating here on different agreement possibilities. Instead, the payoffs listed are derived from enjoying the benefits of a fixed Pareto-superior alternative sooner. To highlight the coordination dilemma facing international actors in altering institutions, therefore, I assume that the superiority of revising institutional structures relative to the status quo is held *constant* in the coordination versus the non-coordination scenarios. Thus, the only difference between acting at T_1 and T_∞ is the possibility of seizing anticipated benefits sooner. States can, for example, enjoy the benefits of *the same* mutually beneficial emission reductions agreement sooner. When incentives exist to update institutions, states receive their largest payoff from acting swiftly.[12]

In the stag hunt game, there are two Nash equilibria: mutual cooperation and mutual defection. The cooperative solution is Pareto-superior, with each side receiving their highest possible payoff. If actors can concert their expectations on the stag-stag outcome, sides end up with the more efficient result. The decision to coordinate is, however, influenced also by the risk of costly coordination failure. When a player chooses to pursue institutional change, they run the risk that others will choose not to. Players who choose not to coordinate minimize risk. Here, as Brian Skryms notes, "rational players are pulled in one direction by considerations of mutual benefit and in the other by

[11] For a related discussion, see Calvert 1998, 75–78.

[12] Distribution problems do not factor into the temporal coordination aspect of the model, yet, as we will see in the empirical chapters, as well as in my use of negotiation analytic concepts, such as the deal-no-deal balance, distributive conflict remains essential to my analysis of institutional change. Distributive considerations factor in through the assessment of incentives to alter institutions.

considerations of personal risk."[13] This uncertainty makes it difficult to coordinate. Without greater certainty, actors are reluctant to assume heightened transaction costs in pursuit of available joint gains because they cannot be assured that the other side will act in concert with them. And this uncertainty is reciprocal. The uncertainty experienced by side 1 adds to the uncertainty that side 2 experiences, which correspondingly reinforces and magnifies the uncertainty of side 1. It is this dilemma that allows the payoff-dominant equilibrium to remain unrealized in favor of the risk-dominant equilibrium. The inefficient rabbit-rabbit equilibrium can thus prevail. Once the rabbit-hunting expectation is established, no actor has incentive to unilaterally defect from it.[14]

In a two-player game with pre-play communication, this should not be an insurmountable barrier to overcome. Signaling or pre-commitment mechanisms can allow actors to move collectively from the rabbit-rabbit scenario to the more efficient outcome. In large multilateral contexts, however, especially when actors are non-unitary, this can be complicated. Signaling is costlier in such contexts. As the number of relevant actors climbs, all things being equal, the more difficult coordination will be. In the game outlined above, side 2 represents an amalgam or composite of other relevant actors. Thus, in employing the stag hunt here, I model side 1's expectations concerning prospects for coordinated action among a large proportion of the institutional population. The higher the degree of uncertainty associated with the behavior of other relevant players, the more attractive persisting at the existing equilibrium will be. This explains the status quo bias observed in many international institutions.

Although international organizations can reduce uncertainty, including through moderating informational asymmetries, they are limited in their capacity to produce such policy-relevant information.[15] Priorities have to be established in the production of information, often following the current preoccupations of member states. While well-functioning and well-resourced international organizations can contribute to reducing uncertainty, therefore, their capacity to respond

[13] Skryms 2001.

[14] There are real opportunity costs here. Costs are proportionate to the length of time actors expect to persist at inferior equilibria, multiplied by the degree of the difference between available institutional alternatives and the status quo.

[15] Abbott and Snidal 1998.

efficiently to progressively emerging social problems is limited. In addition, despite the contribution of international organizations to easing collective decision-making, their size, decision rules, and the heterogeneity of preferences among members can hinder smooth adaptation.[16]

The coordination challenge inherent in the basic stag hunt game described above may be complicated further if we model the impact of incomplete information about the payoffs of other players.[17] Indeed, coordination is the principal means through which more complete information on payoffs is amassed. This entails the possibility that no gains will be available, in part because others assess available incentives differently. While they clarify the nature and intensity of their own preferences through analytical investments, uncertainty is higher concerning the true preferences of other players.[18] From the perspective of side 1, therefore, it assesses its own payoffs to be $A > B \geq C > D$, thus favoring swift coordination. Its evaluation of side 2's preferences is less certain. Here, side 1 faces the prospect of coordination with the following assessment of side 2's payoff structure:

$W > X \geq Y > Z$, with probability p
$Y > X > Z \geq W$, with probability $1 - p$

Here, p is a measure of convergent expectations. This uncertainty is reciprocal. In a world of incomplete information and costly signaling, side 2 – a composite of other relevant actors in the system – faces similar uncertainty about the true preferences of side 1.

In the model presented, actors are receptive to indications about the true payoff structure of the other side(s) concerning the initiation of institutional change processes. It is here where a temporal convergence of expectations matters so greatly. For example, the emergence of a focal time frame could lead them to revise their assessment of the p value from 0.4 to 0.9, for example, thus heightening greatly their willingness to assume the risks associated with moving to alter

16 Lundgren et al. 2018.

17 Game theorists have done extensive work modeling games with incomplete information, including the stag hunt. See, for example, Carlsson and van Damme 1993.

18 This approach follows the work of John C. Harsanyi in modeling the impact of uncertainty in stag hunt games. I would like to thank Dane Rowlands for helpful discussions in operationalizing this extension to my model. See: Harsanyi 1973.

institutions. Conversely, divergent expectations owing to a lack of clear temporal landmarks entail high risks of coordination failure. In such settings, actors are more reluctant to pursue change, making the institutional status quo more attractive. In these contexts, change is less likely.

Actors can, moreover, be understood as boundedly rational in the sense that they face cognitive limitations and time constraints.[19] They are bounded also in one further, yet fundamental, respect: In addition to incomplete assessments of all agreement options, explaining deviation from a strict "efficiency criterion," they do not continuously scrutinize their choices across time.[20] For Herbert A. Simon, the main theorist of bounded rationality, individuals regularly deviate from the efficiency criterion because they fail to consider every possible behavioral option. Acting on force of habit or reflex, decisions are not always self-conscious, deliberate, or rational. Satisficing – a theme that has been used widely in political science literature – can apply both to a lack of optimization among agreement possibilities *and* to the slow rate of grasping those possibilities. Simon's model can thus be extended to include "temporal" satisficing, where actors do not evaluate continuously their choices. Temporal satisficing may explain lags in actors' recognition that extant institutional arrangements are not optimal.

How Does Change Occur?

Implicit in the stag hunt game is the presence of incentives to coordinate. That is, the individual and collective benefits of coordination must exceed those available through independent action. Yet coordination often fails, leaving actors with inferior outcomes. This scenario is analogous to the widely recognized, persistent dysfunction seen in many international institutions today. Two main factors will drive individuals to act cooperatively. First, there must be incentives available for updating an institution. That is, the larger the degree of perceived superiority between the best available institutional alternative and the status quo, the greater the incentive to alter institutions. Second, expectations concerning the likelihood that other actors will

[19] For a treatment of bounded rationality in institutional change efforts, see: Jupille et al. 2013.

[20] Simon 1961, 14, 40.

behave in a cooperative fashion are highly important for achieving coordination. When an actor believes that others will hunt stag – or, in this case, pursue institutional change – they will do so. I will discuss each of these variables in turn.

There must be a perceived incentive to alter institutions. Incentives can be material or immaterial. They can be a consequence of discrepancies between the distribution of influence within an institution, as embodied in its rules and procedures, and the wider distribution of capabilities within the international system. Outdated rules and procedures may not be an efficient response to emerging cooperation problems or give rise to governance gaps. Incentives may also be immaterial, reflecting changes in international norms, ideas, and identities. In this case, rules may not have kept pace with new principled and consensual ideas. The consequences of normative change or governance innovation may not yet be reflected within an institution. As a result of these and other factors, the legitimacy of the institutional status quo declines relative to an alternative ideational or normative framework. All of these developments open up incentives to alter institutional arrangements.

Incentives in institutional bargaining can be analyzed through assessing the deal-versus-no-deal balance among relevant players.[21] Actors have a variety of interests in negotiations, which are impacted by changes in material or normative/ideational conditions.[22] These conditions affect the underlying interests of institutional actors, thus implicating their alternatives to negotiated agreement.[23] The greater the potential value of a deal relative to the costs of impasse, the greater the incentive to update institutions. The relative positions of actors define the outer limits of the Zone of Possible Agreement (ZOPA).[24] A ZOPA can be opened up through capitalizing on available joint gains among actors, including through issue linkages, manipulating the parties/issues under negotiation, and the dovetailing of differences in risk attitudes, forecasts concerning unknown future contingences,

[21] Sebenius et al. 2018, 86–88.

[22] This analysis links also to theories of "ripeness," where a costly stalemate in negotiations can create conditions favorable to agreement. Consistent with the work of Dean G. Pruitt on "readiness theory," I do not assume that available incentives lead inevitably to agreement. See: Pruitt 1997. On Ripeness, see: Zartman 1996.

[23] Fisher et al. 1991.

[24] Raiffa 1982, 45.

and time preferences. When these positions do not overlap (i.e. some actors see no deal as superior to any possible agreement), there is an adverse deal-no-deal balance and actors do not have an incentive to alter institutional arrangements. In such a circumstance, substantial change cannot occur.

It is here that distributive conflict and power politics factor into the equation. Strong alternatives to negotiated agreement provide the basis for the reservation points that define the outer limits of the ZOPA. Actors' "go-it-alone power" can be used to manipulate agreement possibilities and incentives to alter institutions, thus bringing institutional equilibria into closer alignment with structural conditions within the wider international system.[25] Issue linkages can be used to claim, rather than create value. Agreement among multiple possible cooperative equilibria "along the Pareto-frontier" can have sharp distributive dimensions.[26] As we will see in the ensuing chapters, hard bargaining and strategic competition factor heavily into the model via this variable. For those with an appetite for a bit of "blood" in their institutional analysis, this is where it is found.

Actors must also believe that others will behave in a cooperative manner by moving collectively to hunt stag or, as it were, update institutional arrangements. Total certainty in this regard does not exist. Rather, players must decide to act in the face of varying degrees of uncertainty. The decision to hunt stag manifests itself in a probabilistic estimation of the likelihood that other international actors will move to negotiate changes to existing institutional arrangements. The most dynamic factor in increasing the estimated likelihood of coordination is the emergence of a temporal convergence of expectation. Convergent expectations provide a definite time frame for institutional bargaining. A temporal convergence of expectations is the emergence of intersubjectively shared expectations of coordinated choice in time.[27] There is a certain hall of mirrors quality to this. The action of each actor is shaped by their expectations concerning the actions of others; which

[25] Gruber 2000. [26] Krasner 1991.

[27] I want to thank Duncan Snidal for his collaboration in this definition. Relatedly, in typically spare fashion, Thomas Schelling defines convergent expectations as "the shared expectation of an outcome." For Schelling, the "outcome" is coordinated choice. Schelling also gives weight to the importance of stability in defining convergent expectations. Schelling does not, however, seek to define temporal forms of convergent expectations. See: Schelling 1958, 253–254.

are, in turn, predicated on action expectations for the first actor almost *ad infinitum*. As we shall see, such a process can be set in motion by the emergence of a focal time frame. Deals struck and concessions granted in the context of convergent expectations are thus made with a final negotiated resolution in mind, thus easing the temporal variant of the negotiator's dilemma described in Chapter 1. For the foreseeable future, actors do not anticipate granting their negotiating counterparts another bite at the apple.

Concretely, convergent expectations normally become associated with a specific outlet for negotiations. In the case of an international crisis or exogenous shock, cooperative problems that come to light as a result of that event are blended definitively and unambiguously with an institutional forum and a specific timeline when expectations converge. During the global financial crisis, for example, the November 2008 G20 Washington Summit emerged as a high-level focus for addressing a pressing economic situation. By contrast, during the 2020 novel coronavirus outbreak, no one institutional venue or timeline emerged for orchestrating a global response to the crisis. Coordination was hindered by epidemiological dynamics, including a gradual and uneven initial spread of the virus as well as variation in country impacts.[28] There was a temporal *divergence* of expectations. Coordination can also be achieved through other means. In the case of a Temporal Focal Point (TFP), convergent expectations are realized when international actors associate an existing institutional problem with a focal time frame. The problem of mass atrocity crimes became tied explicitly to the approaching sixtieth anniversary session of the United Nations in 2005, adding impetus to the endorsement by the UN General Assembly of the Responsibility to Protect doctrine. In this case, the explicit link between an important anniversary and a pressing problem provided a point of focus for intergovernmental negotiations.

Convergent expectations are, in this conceptualization, an emergent phenomenon. Their power does not rest in the actions or intentions of any one individual. They cannot be harnessed or controlled, for example. They are a social dynamic that emerges among groups of actors. Their salient properties are thus collective, rather than

[28] Patrick 2020.

Table 2.2 *Convergent expectations and change incentives*

	Convergent expectations	Divergent expectations
Incentives	Major institutional change	Suboptimal institutional status quo
No incentives	No substantive agreement, status quo persists	Self-enforcing institutional status quo

individual. Just as schools of fish or flocks of birds feature distinct behavioral characteristics, including regularized patterns of choice irreducible to the decisions of individuals, the rules governing international actors must be observed and understood collectively. No one is driving the bus, as it were. The actions of individual agents are less important than those of the whole. It is not generally possible for even well-informed actors to have a comprehensive understanding of behavior unfolding at the macro level.

These conditions – available incentives and a temporal convergence of expectations – are individually necessary and collectively sufficient to produce change of great speed, scope, and depth. When incentives exist, but actors face temporal coordination dilemmas, a suboptimal institutional status quo will persist. This is a ubiquitous circumstance in global politics today. Convergent expectations without incentives to alter institutions also cannot result in significant change. Incentives to alter institutions are the raw materials for change. Without such materials, actors have no basis for revising institutional arrangements in a substantial way. In this circumstance, states may produce joint statements lacking substance or significance to mark associated international meetings or may agree on smaller institutional amendments, but cannot fundamentally alter the institution. More often, talks will break down. When combined with incentives, however, convergent expectations will lead international actors to realize change. Table 2.2 summarizes the impact of these two conditions.

These independent variables produce two main patterns of punctuated institutional change (summarized in Table 2.3). The first, highly familiar pattern in world politics occurs when convergent expectations are achieved roughly concurrent with the emergence of incentives to

Table 2.3 *Triggers for convergent expectations in time*

	Exogenous shocks	Temporal focal points
Temporal relationship to change incentives	Concurrent	Dissociated
Associated with change incentives that emerge	Suddenly	Gradually
Common sources	Wars, depressions	Demonstration effects, anniversaries

alter institutions.[29] This is the exogenous shock pattern. In addition to being associated with the breakdown of the existing cooperative order, shocks are – for want of a better word – shocking. Wars, depressions, or other transformative social events have a way of focusing minds on the state of global relations. Thus, in addition to creating incentives to alter institutional arrangements, exogenous shocks generate a widespread, urgent focus among relevant players on the revision of institutional norms, rules, and procedures. This intense focus is of great significance for the reestablishment of cooperative equilibria in the wake of a shock. The intellectual and analytical resources devoted to planning peace after the World Wars, for example, were monumental. Actors inside and outside of government focused intensively on the shape of the postwar order, making heavy investments to exercise influence during these phases of institutional flux.[30] With shocks, the scale of change should be proportionate with what is needed to reestablish a cooperative equilibrium.

Shocks have been such an important generator of change in world affairs that international relations scholars have overlooked the extent to which shocks also generate coordination. Thus, in assessing change associated with an exogenous shock, there is risk of omitted variable bias. Shocks simultaneously produce both significant change in

[29] In reality, coordination quickly follows the emergence of change incentives. The institutional response is, however, comparatively seamless and, consequently, there is a minimal lag between the emergence of incentives and coordination so as to appear concurrent.

[30] Chapnick 2005; MacMillan 2003; Schlesinger 2004.

international conditions, the basis for new equilibria, *and* a temporal focus that facilitates coordination. The visibility and prominence of shocks helps to explain the temporal proximity of change to the conditions that helped produce it. Indeed, it is the simultaneous emergence of incentives and the need for coordination that gives exogenous shocks their bite. Also significant in dissecting the impact of shocks on institutions is the degree of proportionality between the scale of change seen and the magnitude of the shock. A strict application of the causal logic of exogenous shocks would suggest that institutional adaptation should be proportionate to what is needed to reestablish equilibrium. Change in excess of this could owe to the coordination effects derived from the shock and should be analyzed carefully when attributing large-scale change to shocks.

A second pathway to significant institutional change is when the emergence of incentives to alter institutions and convergent expectations arrive separately. That is, they are dissociated temporally. Actors have an incentive to reform institutions but remain mired in the less risky institutional status quo. Here, frustration may build as institutions underperform and as their legitimacy erodes, yet states cannot generate the coordinative impetus to address the situation. The persistent dysfunction seen in many international bodies suggests that this is a common scenario in institutional life. Much of the coordination dilemma facing states owes to the inherent status quo bias of institutions and the fact that many incentives to revise institutions emerge gradually through exogenous shifts or reactive sequences. Yet, the arrival of convergent expectations can rapidly change the equation. A temporal focus leads actors to heighten their estimation of the likelihood of coordinated institutional change efforts. In this scenario, a wide array of actors make political and analytical investments associated with institutional change processes. Such investments transform the informational and political context, leading to the clarification of preferences and a realization of previously available incentives to revise institutional structures. Previously latent preference structures are made manifest through conscious deliberation, leading to better alignment of negotiating stances on institutional questions with altered global realities.

Change seen in the context of convergent expectations adds to institutional analysis along what Johannes Gerschewski calls the "missing diagonal."[31] The time horizon for the cause of institutional change in these instances can be long, at least in relation to the initial shifts in international conditions that motivated it, and exogenous. While my analysis focuses on large-scale institutional change, convergent expectations might also facilitate smaller-scale adjustments, consistent with literature on more gradual forms of institutional change. When the motivations for change are endogenous in nature, such as following the development of new principled or causal ideas, furthermore, we might see change on a short time horizon. The analysis of convergent expectations in time, therefore, helps us to understand understudied patterns of institutional continuity and transformation.

Temporal Focal Points

A temporal convergence of expectations plays an essential role in actors' capacity to realize institutional change. In the case of exogenous shocks, such a convergence is part of the package. By virtue of the shock being a shock, a temporal focus is all but assured. In cases where incentives accumulate gradually, however, reaching convergent expectations can be much more difficult. It is here that Temporal Focal Points are of special relevance.

In his *Strategy of Conflict*, Thomas Schelling provides a memorable example of a tacit coordination problem, asking how one might hope to meet someone in New York City if they could not communicate with them.[32] The majority of his respondents were able to concert their expectations on Grand Central Station's information booth at noon. According to Schelling, Grand Central Station served as a coordination point because of its conspicuousness and exceptionality relative to other possibilities. These traits made the station a point of focus for his respondents, enabling a convergence of expectations.

Schelling's analogy has gained wide attention in the literature. Most of it has, correctly, focused on his main idea: In settings characterized by multiple equilibria, conspicuous focal points have great coordinative power. One aspect of the story that is often overlooked, however, is that Schelling's respondents were coordinating on two dimensions: space and time. They met at Grand Central Station *at noon*. A further dimension remains implicit in Schelling's discussion: the *date* of

[31] Gerschewski 2021. [32] Schelling 1960, 55–56.

coordination. While coordination on multiple dimensions may be unimportant in certain settings, many social contexts involve just this challenge. Even if actors can concert their expectations on one dimension, divergent expectations on a second or third dimension can lead to coordination failure.

Could the rise of conspicuous events – such as high-profile anniversaries or visible crises – serve as a coordination mechanism for states in a manner analogous to Schelling's concept of focal points? In the context of multiple equilibria, Schelling discusses "the intrinsic magnetism of particular outcomes, especially those that enjoy prominence, uniqueness, simplicity, precedent, or some rationale that makes them qualitatively differentiable from the continuum of possible alternatives."[33] Just as political actors face a continuum of qualitatively undifferentiable bargaining solutions, they also face a temporal continuum of qualitatively undifferentiable bargaining timelines along which they must coordinate their behavioral expectations.

Temporal Focal Points are defined by three main features.[34] First, they occur within a discrete, identifiable time frame. A scarcity of temporal signposts and perceived continuities in time are the enemies of purposeful negotiation.[35] The perception of discontinuity or nonlinearity along the temporal continuum – the passing of a distinct phase or the qualitative differentiation of a period in time – is vital to achieving convergent expectations. Perceptions of the focal time frame must be broadly shared. In practice, this means that TFPs usually become associated with a particular negotiating process, which gives definition to the bargaining timeline. Similar to Schelling's focal points, except along the temporal dimension, TFPs enable coordination in contexts of multiple equilibria.

The precise manner in which a TFP brings a definite time frame to institutional settings is an empirical question. Cases may differ

33 Schelling 1960, 70.

34 Temporal Focal Points are, of course, a stylized concept. They are articulated here in their ideal typical form. The exogenous origins of an event always have some aspect of human agency associated with them. In practice, some temporal foci are likely to require a higher degree of agency and entrepreneurship than others. TFPs are likely not possible without some agency. No focal point is inevitable. At the same time, certain coordination points have an almost intrinsic magnetism that owes relatively little to agency and entrepreneurship.

35 Gersick 1988. See also: Gersick 1989; Lim and Murnighan 1994.

markedly. Actors might react swiftly to a crisis or capitalize on the coordinative impetus that a visible event brings to a long-standing problem.[36] The arrival of an important anniversary or a high-level political gathering may act as a stimulant for negotiations, enabling actors to reach agreement within the ambit of a particular forum. The empirical particulars are less important than the observation that certain points along the temporal continuum – because of their conspicuousness – can prompt international actors to behave as if there is a major temporal discontinuity in bargaining. States negotiate with great intensity to realize agreements before self-imposed deadlines or suddenly make major concessions to "salvage" talks that risk an impasse.

Second, focal points are unique. There must be something unusual and important about what would otherwise be an indistinct point along the temporal continuum. If there exists more than one equally salient time frame to focus the attention of actors at any one point of time, a divergence of expectations may result. Uniqueness avoids ambiguousness and brings prominence. Schelling gives the example of two people seeking to meet in a small town by referencing a map: "a map with many houses and a single crossroads sends people to the crossroads, while one with many crossroads and a single house sends most of them to the house."[37] As we will see in Chapter 5, for example, ambiguity over the follow-up to the Brundtland Commission report contributed to a divergence of expectations and coordination failure. TFPs are the temporal equivalent to the town's one house or crossroads in Schelling's analogy. The greater a moment's claim to extraordinariness, the stronger its pull.

Third, Temporal Focal Points are highly conspicuous. Like focal solutions to bargaining problems, TFPs must be prominent, attracting the attention of actors. Agreements during focal phases are the "obvious" or "natural" time to find a settlement on a pressing issue.[38]

36 One example of this was seen during the Kosovo crisis in 1999. Serbian officials, conducting a campaign of ethnic cleansing and fearful of a North Atlantic Treaty Organization (NATO) intervention, quipped that "a village a day keeps NATO away." While the situation in Kosovo grew progressively worse for ethnic Albanians, Serbian forces sought to prevent the rise of a visible crisis that could enable coordination among NATO countries. A well-publicized atrocity in February 1999 in the village of Račak changed this equation. See: Gellman 1999; Manulak 2009.

37 Schelling 1960, 58. 38 Schelling 1960, 68–69.

Although important decisions could have just as conceivably been taken in 1991 or 1993, for example, the 1992 Earth Summit in Rio de Janeiro was widely seen as the obvious time to reach agreement on global environmental problems. Since no other juncture in that sphere of institutional activity stood out in a comparable manner, Rio served as a point of focus for relevant actors. Because of their conspicuousness, TFPs tend to have an elevated profile and generate high-level political participation. While major conferences and negotiations are common sources of TFPs, routine meetings can also be imbued with conspicuousness.

Given the preceding analysis on the role of exogenous shifts and the importance of convergent expectations in institutional life, this assessment also suggests that crises commonly touted, *ex post*, as exogenous shocks can often be fueled by long-standing institutional deficiencies. Even when crises appear to cause the breakdown of cooperative equilibria, the real work has often been done by accumulated exogenous shifts and the arrival of a temporal point of focus. As Margaret MacMillan notes, often shocks "do not come out of the clear blue sky," but are instead produced by a series of gradual causes, such as the progressive weakening of restraints imposed by prevailing norms and institutions.[39] While apparently abrupt changes highlight that past behavior is no longer an equilibrium, triggering events may not themselves cause transformations. An equally likely possibility is that a conspicuous event, such as a crisis, provided a coordination mechanism for actors to realize accumulated cooperative possibilities thus overcoming persistent institutional deficiencies.

No TFP is inevitable. All focal points exist on a spectrum between purely exogenous and almost wholly endogenous.[40] For some focal points, their coordinative power owes almost exclusively to the institution or subject matter in question. The temporal landmark is almost completely "found," external to implicated actors, analogous to the rising and setting of the sun for marking time.[41] In other instances, the coordinative potential of a focal point is less obvious, less widely

[39] MacMillan 2020, 14.

[40] Here, my analysis allows for a greater degree of endogeneity in TFPs than recognized in my 2020 article in *Review of International Organizations*, while maintaining that TFPs must have an exogenous component. See: Manulak 2020.

[41] The helpful terminology of "found" versus "constructed" marks and measures comes from Nomi Claire Lazar 2019, 21.

recognized, and thus has to be more heavily "constructed." It is not uncommon to see entrepreneurial actors implicated in change associated with TFPs in these cases.[42] TFPs are thus *crystallized* by institutional actors, who enhance the conspicuousness of focal points. Crystallizers tend to be well-positioned within actor networks, serving as connectors in institutional life. In the language of social network analysis, they enjoy high degree or betweenness centrality, possessing a large number of direct connections or links to key, otherwise weakly connected segments of the institutional population.[43] They may also act as points of linkage between discrete networks, such as scientific, business, social, or policy communities. In this sense, crystallizers can serve as multiplex nodes in institutional affairs or network "switchers."[44] This formulation is not altogether dissimilar from that of Wesley Widmaier and colleagues, who argue that the impact of exogenously generated events is enhanced greatly by actors seeking to advance their own agendas.[45] Yet focal points are not "endogenous constructions." Crystallizers do not *create* TFPs without having something already to work with. They amplify some temporal mark or measure that is already there. The impact of such actors is related inversely to the degree to which the TFP provides a compelling coordinative logic unrelated to the aims of any subset of the institutional population. The more intrinsically compelling the TFP, the less constructed it is.[46]

The analysis of temporal foci is supported by research in related disciplines.[47] In the field of management science, Connie J. G. Gersick's

42 For an approach to analyzing the interests of agents interacting with institutional structures, see: Büthe 2016.

43 On different measures of centrality, see: Scott 2017. While it is possible that focal points might spread through the network from less propitiously located actors, this is much less likely.

44 For a discussion of multiplex nodes, see: Avant and Westerwinter 2016. On network "switchers," see: Castells 2012, 8–9.

45 Widmaier et al. 2007, 747–759.

46 Andrew Hom provides another distinction between "active" and "passive" means of timing. The better established the timing mechanism, through precedent and other means, the more passive is coordination achieved through it. See: Hom 2020, 36.

47 In their simulations, Stephen Ghee-Soon Lim and J. Keith Murnighan found strong evidence that temporal-pacing was influenced by deadlines. The influence of the deadline was felt, in their studies, some time after the midpoint. Their work concurs broadly with "Parkinson's law," which states that there is an elastic element to time: "work expands so as to fill the time available for its completion." Thomas Sandholm and Nir Vulkan assessed the impact of

groundbreaking studies on the effect of transition points on group task completion show that groups regularly use temporal signposts to govern their behavior.[48] In Gersick's experiment groups, there was a period of inertial drift that lasted through half of the predetermined timeline for task-completion. After that, each group, "underwent a major transition. In a concentrated burst of changes, groups dropped old patterns, reengaged with outside supervisors, adopted new perspectives on their work, and made dramatic progress."[49] Even within the post-transition phase of work, groups' rate of task completion accelerated greatly as they neared deadlines.[50] Ron E. Hassner's provocative analysis of "sacred time" assesses the strategic implications of military operations planned to coincide with dates of religious significance.[51] The use of sacred dates demonstrates that time is not homogeneous in the behavior of many military actors. An important factor in his analysis is the coordinative potential of military campaigns associated with certain points along the temporal continuum. For instance, initiating conflict on sacred days can be used to bolster the motivation of troops. Hassner notes that: "War initiators can also expect civilians, observers, and third parties who share their interpretation of the date's religious symbolism to support their cause and mobilize in its defense."[52] Hassner finds that days of religious salience are thus disproportionately likely to see conflict initiation.

Because time is a continuous variable, actors find it difficult to concert their expectations along the temporal continuum. This process is eased, however, by the emergence of prominent temporal signposts, such as deadlines, midpoints, or sacred dates. Such dates can help produce convergent expectations, activating the emergent properties at the core of decentralized coordination. Conspicuous phases along the temporal continuum are disproportionately likely to produce behavioral change and coordination because actors expect that the attention of other actors will be drawn to some salient characteristic

deadlines on bargaining through a series of computer simulations. In bargaining, when firm deadlines are a form of private information – not an unreasonable assumption in talks where parties approach bargaining with different degrees of urgency – making first offers can be seen as a sign of weakness. Even when actors stand to gain from agreement, the rational choice is to attempt to outwait one's opponent. See: Lim and Murnighan 1994; Parkinson 1980, 13; Sandholm and Vulkan 1999.

48 Gersick 1988. See also: Gersick 1989.
49 Gersick 1988, 16.
50 Gersick 1988, 18.
51 Hassner 2011.
52 Hassner 2011, 504

of that moment in time. Midpoints and approaching deadlines in task completion cause actors to notice the fact that time is running out. Sacred days have connotations that make certain religious dates stand out from the continuum of days in which a conflict may be initiated. These are all temporal examples of tacit coordination analogous to Schelling's Grand Central Station example. Importantly, in each of these cases, actors were coordinating along more than one dimension.

Convergent expectations generate institutional transformations through a belated recognition among relevant players of previously overlooked shifts in the international environment. Heightened political and analytical investments lead actors to recognize an interest in revising cooperation structures and thus to act to realize available incentives. The mobilization of latent coalitions may also shape the political and informational conditions facing actors. Windows for persuading key players through the provision of information may be opened by convergent temporal expectations. New agreement possibilities come into view, allowing actors to craft issue linkages or dovetail differences in risk attitudes, time preferences, and forecasts concerning unknown contingencies. In so doing, TFPs help to generate a temporal convergence of expectations.

Observing Temporal Coordination and Institutional Change

I take a dual approach to theory testing (summarized in Table 2.4). The first is at the level of comparative statics. Here, I track key outcomes predicted by my framework across cases. The preceding analysis has presented two conditions that are individually necessary and collectively sufficient for large-scale institutional change: the presence of incentives to alter institutions and a temporal convergence of expectations. Incentives can be material or immaterial, manifesting themselves in the availability of joint gains through alternative institutional arrangements. Convergent expectations work through increasing the assessed likelihood that other international actors will coordinate, leading to a clarification of preferences and influencing bargaining processes. TFPs can help to precipitate such a convergence. These variables give rise to two patterns of institutional change: one that follows exogenous shocks and one that occurs when incentives to reform institutions accumulate more gradually. Unlike shocks, where both variables are simultaneously present, actors face coordination

Table 2.4 *Levels of theory-testing*

Level of theory-testing	Observable implications
Outcome-based analysis	• Institutional change when both independent variables are present: change incentives and a temporal convergence of expectations
Process-based, Within-case assessment	• Significant political and analytical investments follow, rather than precede, convergent expectations; • Bargaining strategies: integrative bargaining sequenced in relation to focal time frame; • Temporal-pacing: increased frequency or length of meetings, creation of working groups, transition points.

dilemmas when incentives accumulate gradually and are reluctant to invest in change processes until they expect others to do so. In this circumstance, the span in time between change in international conditions and consequent institutional adaptation can be significant. This can change suddenly when expectations converge, often triggered by the arrival of a Temporal Focal Point.

The theory can be tested by assessing the presence or absence of both conditions in instances of large-scale institutional change. Following Thomas Rixen and Lora Anne Viola, I define large-scale institutional change as alterations in institutional norms, rules, and procedures of great speed, scope, and depth.[53] Change of great speed, scope, and depth has a non-incremental character, transforming cooperation within that institution at a fundamental level.[54] Such change can take the form of both a branching tree style of institutional development and of punctuated, step change.[55] In these instances, actors make a

[53] For Rixen and Viola, speed refers to the time it takes for a given institutional change to occur. Scope and depth are two measures of the magnitude of change. Scope focuses on the number of institutional features affected by change. Depth refers to the degree to which key institutional features change. See: Rixen et al. 2016a, 18–20.

[54] On more gradual forms of change, see: Mahoney and Thelen 2010; Young 2010.

[55] Although Johannes Gerschewski (2021, 222–224) views branching tree modes of change as a form of gradualism, I maintain that this depends on the scope and depth of that change.

deliberate choice to alter patterns of cooperation.[56] Such change should occur, according to the theoretical framework presented in this book, when both of these conditions are present and should not occur when one or both of them are absent. If the international context evolves gradually, creating unrealized incentives to alter institutions, then we would expect the timing of change to be influenced chiefly by the attainment of convergent expectations. From this perspective, incentives take the form of a *permissive* condition for institutional change. They are necessary for the realization of change, but often remain in the background awaiting a shift in other conditions. Convergent expectations are regarded as a *triggering* condition for change. In the presence of the permissive condition, they serve as the immediate stimulant for the observed phenomenon. Convergent expectations may be facilitated by a TFP, as well as other stimulants, such as visible crises. Thus, the finding that the timing of institutional change is associated more closely with the arrival of convergent expectations than with the presence of incentives would be an observable implication of the theoretical framework presented here.[57]

The second level of theory testing, undertaken mainly in Chapters 3–6, is at the process level. Through within-case comparisons, we are able to generate further observable implications. A central implication of the coordination dilemmas highlighted in this book, particularly in cases where change incentives accumulate progressively, is a reluctance among actors to make major political and analytical investments in institutional change processes until expectations converge. Such investments fundamentally alter informational and political conditions, allowing actors to clarify the intensity and nature of their preferences. A temporal focus also opens the door to major shifts in bargaining behavior. A key implication of temporal coordination dilemmas therefore is evidence of temporal-pacing among actors in bargaining and in making vital investments in change processes. The arrival of convergent expectations will lead actors to increase their probabilistic assessments of the likelihood of coordinated efforts to alter institutions, motivating them to expend scarce assets to influence institutional bargaining. There is a sudden willingness to incur heightened

[56] In contrast, gradual institutional change through layering or drift can have a less deliberate and deliberative quality, with successive incremental choices having a large cumulative impact.

[57] For a related assessment, see: Soifer 2012.

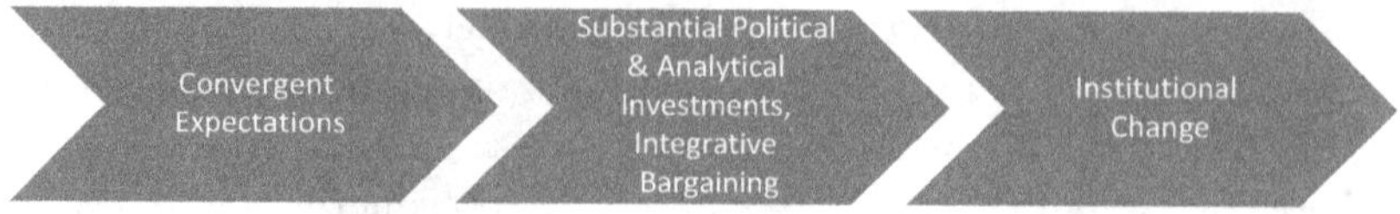

Figure 2.1 The temporal coordination model of change

transaction costs. As a result, *after* expectations are aligned, we expect an increase in such investments from an array of national and international players seeking to act within the focal time frame. This process is illustrated in Figure 2.1. Before a temporal convergence of expectations, states chase rabbits alone. After, they hunt stag together.

Conversely, other theoretical treatments expect coordination to be a by-product of changes in international conditions or norms/ideas. An understanding of the shifting global environment is produced by ongoing political and analytical activities by national and international organizations. Such activities may peak as actors seek to decide whether to coordinate institutional bargaining to address changed conditions through revising institutions. We would therefore expect that substantial investments would precede and, indeed, motivate convergent expectations. Thus, rather than such investments being made in response to an emergent expectational convergence, political and analytical investments should in large part occur *before* the emergence of convergent expectations. Successful norm entrepreneurs, for instance, progressively build support for their normative agenda, triggering a cascade that provokes convergent expectations in support of change.[58] Similarly, new principled ideas can mount or new information can demonstrate that institutions are suboptimal, leading to coordination and institutional change. In all of these cases, it is the growth of information and a groundswell of political support that spurs international actors into coordination that leads to change in institutions. This process is illustrated in Figure 2.2. There is thus an implicit tipping point logic underpinning these accounts. Convergent expectations are a corollary of advances in information, analysis, and entrepreneurial activities.

If the model presented in this analysis is correct, we would expect to see the counterintuitive pattern that, rather than spiking *before* expectations converge, consequential information and analysis of

[58] Finnemore and Sikkink 1998, 900–901.

Figure 2.2 Conventional models of institutional change

institutional problems proliferates in large part *after* expectations converge.[59] Responding to the prospect of coordination, there is sharp increase in the volume of reports, publications, or other analyses produced in relevant government departments, nongovernmental organizations, think tanks, and international organizations. These products are drafted to coincide with an available bargaining window. The quality of information and analyses produced once temporal expectations converge may matter more than the quantity. The temporal coordination framework would expect that a significant share of "big ideas" and recognition of changes to the "state of the world" emerges during focal time frames. Ideas arriving after expectations converge are thus at the "strategic" level as well as at the "tactical" level. Other approaches, on the other hand, would expect that key information and ideational breakthroughs would *precede* a convergence of expectations in time.

Two other process-level observable implications can be specified. Tied closely with these informational investments are activities that affect the balance of political support for institutions. Such activities can involve the creation of special working groups across government departments, task forces, or the convening of committee hearings. Other observable indicators include spikes in the number of governmental and nongovernmental institutional diagnostics, the introduction of new reform proposals, and an increase in media reporting on institutional questions. Actors build connections within related networks and seek to forge supportive coalitions with a widening array

[59] Admittedly, this account is somewhat stylized. Some – perhaps many – international actors make ongoing investments in an effort to trigger coordination, even when such efforts have minimal chances of success. As with all models, therefore, the one outlined here simplifies reality to make a more fundamental analytical point. It captures an important dynamic at play when a sizable proportion of actors do indeed withhold a significant proportion of the possible political and analytical investments they make until a phase of institutional bargaining seems likely.

of actors. Thus, political forces that were previously latent, likely because it was difficult and resource-intensive to locate potential coalition partners, now emerge as an overt political force.

In addition, in negotiations we observe bargaining behavior that conforms more fully to the logic of convergent expectations than to one of responding to the availability of change incentives. For example, negotiations may be explicitly launched, accelerated, or otherwise predicated on meeting a focal timeline. States may, for instance, accelerate negotiations to ensure consideration within the ambit of a focal phase. An increased frequency or tempo of meetings and the willingness to forward incomplete proposals to meetings are indicators that states aim to complete agreement within a focal time frame. Conversely, states may stagger serious bargaining or even table otherwise incomplete cooperative frameworks to enable final agreement on their preferred timeline. If no consequential agreement is reached, states may employ face-saving measures to give the appearance of progress. Indicators of temporal-pacing include the frequency of meetings, the creation of working groups or task forces, and evidence of transition points in talks.[60] All of these observable factors indicate that the completion of agreements may be affected by a temporal convergence of expectations. Importantly, if the availability of incentives is the driving force in bargaining, then this pattern of behavior makes little sense.

In this respect, temporal coordination dilemmas will often manifest themselves in relation to actors' bargaining strategies. Rather than acknowledging an interest in agreement, actors will bargain cautiously or defensively when expectations diverge. They may put off negotiations altogether or will make few concessions to help bring about agreement. Instead of communicating openly in a joint search for positive-sum solutions, posturing and inflexible negotiation tactics will dominate. Such positional bargaining strategies will persist until the emergence of a focal time frame. When coordination dilemmas are eased, enabling a temporal convergence of expectations, actors will

[60] Some of these indicators were employed in Connie Gersick's group work simulations. See: Gersick 1989, 12–15. If one has access to a fairly complete archival record, including documents from the period preceding and following the principal negotiation stage, it may be possible to gather quantitative data on the frequency and length of meetings, as well as on the timing of key transition points in talks. While I observe some of these indicators in this book, they are assessed primarily using qualitative methods.

suddenly shift gears and engage seriously. Negotiations, which lacked impetus, suddenly find their direction. Transitioning to an integrative bargaining approach, sides will sequence concessions progressively – and reciprocally – to move toward their true bottom line in talks. Such a transition occurred in the context of the 2015 Paris climate agreement, where states moved "away from competition and toward shared winning," thus "unlock[ing] the door to the global agreement."[61] While this process may seem on the surface to pertain more to the presence or absence of incentives or a ZOPA, it is a form of temporal pacing consistent with the resolution of the temporal variant of the negotiator's dilemma discussed in Chapter 1.

Studying Temporal Coordination over Time

The process-based, within-case studies contained in this book will be analyzed through the qualitative method of structured, focused comparison, which identifies and systematically compares variables of interest across case studies, standardizing data collection.[62] Detailed qualitative research is well-placed to discern the influence of subtle interactive effects, to evaluate competing explanations systematically, and to address problems of equifinality.[63] A longitudinal study can also determine whether slow-moving processes are responsible for institutional change in these cases.[64]

This structured, focused comparison will be combined with a "before-after" research design.[65] Such an approach involves dividing the case study into several sub-cases. This approach has the virtue of

[61] Christian Figueres and Tom Rivett-Carnac (2020, 62–63) discuss the transition to integrative bargaining, in the context of what I term a Temporal Focal Point, the Paris climate conference, as moving to a mindset of abundance:

> The new understanding established that reducing emissions is indeed a responsibility for every nation, for its own enlightened self-interest *as well as* for the planet as a whole. The mindset shift and associated new language in the text – away from competition and toward shared winning, where everyone can gain from a new abundance without impinging on each other – unlocked the door to the global agreement that would be signed in Paris the following year.

While Figueres and Rivett-Carnac do not discuss temporal coordination issues or TFPs explicitly, it is implicit that the shadow of the Paris conference impelled dealmaking and the transition to integrative bargaining in 2014–2015.

[62] George and Bennett 2005, 67–72. [63] Bennett and George 1997, 21–22.

[64] Pierson 2004, 86–92.

[65] For a discussion, see: George and Bennett 2005, 166–167.

achieving a high degree of control in analyzing changes in variables. In particular, it allows me to track shifts in institutional change incentives over time, as well as the presence or absence of convergent expectations. Because key variables remained constant through several of my cases, this research design also allows me to control for potential alternative explanations, including: the presence of specific institutional alternatives and norms, preference aggregation or voting methods within an institution, the bargaining leverage and institutional objectives of negotiating parties, the state of consensual knowledge, and the composition of potential reform coalitions. Such an approach complements analyses at the level of comparative statics.

A particular focus in my process-level investigation will be the analysis of the timing in which incentives to revise institutions emerged relative to the timing of successful coordination. If the arrival of incentives to alter institutions predate concerted intergovernmental decision to act on those incentives, suggesting that a risk-dominant equilibrium persisted longer than might be anticipated by pure forms of rational choice institutionalism, then this approach will demonstrate it. If, on the other hand, international actors are generally able to respond swiftly to shifting international parameters, then this research design will demonstrate that as well. A before-after research design is ideally suited to studying the coordination dilemma imposed by gradual shifts in global conditions and factors that allow states to overcome those challenges.

Another methodological consideration is the risk of covariance between my two independent variables. On balance, the larger the incentive to alter institutions the more likely coordination can be achieved. Similarly, with coordination, the greater the likelihood that the defects of the institutional status quo relative to achievable alternatives will be recognized. Despite this, the lengthy disconnect between changes in international conditions and effective coordination in many institutional settings suggests that there is significant variance between these two conditions, thus minimizing problems of multicollinearity.[66]

Case Selection and Data Sources

The case studies selected have *least-likely* qualities for an analysis of actors' capacity for coordination in institutional affairs.[67] This is

[66] King et al. 1994, 112.
[67] For a discussion of most versus least likely cases, see: Eckstein 1975.

because coordination problems are so severe in UN environmental institutions. The multitude of diverse players active in global environmental politics make it less likely that coordination will occur. At the Stockholm, Nairobi, Rio, Johannesburg, and Rio+20 conferences, for example, more than 100 states participated in talks. Because international environmental issues touch on many spheres of government activity, hundreds of influential subnational actors had to be included in the preparatory processes, including different levels of government and elements of the national bureaucracy. Scientific unions, international media, and nongovernmental organizations were also an essential part of the process. Small, relatively homogeneous areas of cooperation, conversely, are far more likely to see coordination when a salient time frame emerges. The shared cultural or historic experiences of smaller, more homogeneous institutional realms are less likely to suffer from a paucity of coordinative landmarks, including Temporal Focal Points.

The sharp distributive dimensions of North/South relations on the global resource management questions at the heart of international environmental governance, furthermore, means that states had a lot to lose if they were exploited in talks. Like other issue areas that impinge upon states' exercise of sovereignty, economic development, or access to natural resources, distributive concerns are essential to understanding UN environmental politics. The importance that states attached to the development of UN environmental institutions is reflected in the fact that, in addition to the typical diplomatic channels, consideration of institutional change in these case studies involved direct participation at the ministerial or summit level. An evaluation of the archival record belies the not uncommon belief that global environment issues are a relatively congenial corner of international "low politics." Simply put, the cases that I have selected put the framework presented in these pages to a tough test.

At the same time, the case studies selected are, in many instances, *most-likely* ones for rational choice, historical, and sociological institutionalisms. The mid-1960s and mid-1980s were periods characterized by dramatic shifts in underlying conditions that created incentives to revise existing modes of environmental cooperation. Rational choice institutionalism might, thus, predict significant institutional change in the mid-1960s and mid-1980s. To the extent that precise predictions can be specified, historical institutionalists would anticipate reform when institutional arrangements are the least locked-in. On this basis, change in the mid-1960s would have benefited from a weakly entrenched regime context. Historical institutionalists would also expect a transformation, when self-undermining processes had badly weakened existing organizational structures. Sociological institutionalists might expect change when norms and key ideas reached a critical threshold. The periods that environmental ideas shifted most significantly may have been the early 1960s or the late 1980s.[68] By focusing on normative and ideational factors alone, change might be anticipated before the Stockholm conference or around the time of the Brundtland Commission's report. Key alternative explanations, therefore, predict significant change at several junctures assessed here. The expectations of competing theories diverge significantly from those presented in this book.

The insights gained from the case studies contained in the following pages provide strong evidence for the veracity of my framework. The model is tested in an institutional context in which a theory based on coordination would encounter a tough challenge. Given several notable exogenous, endogenous, and ideational transitions in the cases examined, furthermore, one could on the basis of received theory predict institutional change in several instances. Consequently, there is reason to believe that a successful test of my framework would produce generalizations of broad relevance to the study of change in international institutions.[69]

A further methodological challenge of this study is case selection. Drawing causal inferences from the cases discussed above encounters similar methodological limitations as those that were confronted by Theda Skocpol in her study of revolutions: namely that the assessment of failed revolutions is difficult or impossible. A comparison of instances of institutional change would generate selection bias, for example, entailing selection on the dependent variable. To address this issue, I use negative, no change cases – the 1982 Nairobi conference,

[68] It should be noted that Steven Bernstein's account – which emphasizes normative change – would contest this viewpoint. Bernstein argues that the norm complex of liberal environmentalism evolved gradually throughout the period covered in this book until it was institutionalized at the 1992 Rio Earth Summit. See: Bernstein 2001.

[69] For a discussion of the methodological basis for this conclusion, see: George and Bennett 2005, 120–122.

the UN General Assembly's response to the recommendations of the World Commission on Environment and Development in 1987, and the 2002 Johannesburg summit – which resemble Skocpol's use of "moments of revolutionary crisis" as control cases.[70] The four central cases that I study represent major junctures in the history of UN environmental cooperation during which its institutional design came into question. To further mitigate selection bias, I study the intervening years between the focal cases being examined in detail. This approach limits the danger of omitted variable bias and ensures conditional independence.

Data for the case studies has been gathered primarily from archival sources and from elite interviews. The detailed archival documentation was essential to enabling the detailed process-based, within-case, analyses involved in my approach to theory-testing. I have utilized American, Brazilian, British, Canadian, and Swedish diplomatic preparatory documents for the Stockholm conference and for the early operations of UN Environment Programme (UNEP). The documents of the United Nations Archives and Record Management Section in New York and the Rockefeller Archive Center in Sleepy Hollow, New York, were employed to gain further perspective on the Stockholm case. The Maurice F. Strong and Peter S. Thacher papers, held in the Harvard University Environmental Science and Public Policy archives, served as a vital resource on the Stockholm conference, UNEP review processes, the Brundtland Commission, and the Rio Summit. Because scholars have not yet utilized many of the documents that I employ, this study will contribute new insights and knowledge to the record of late twentieth century environmental negotiations and diplomacy. As a supplement to archival research, I have conducted a carefully planned program of interviews with major policy- and decision-makers. These interviews were used to supplement and fill gaps in the document record. Interview subjects were chosen based on the substantive role that they played in the events covered in this examination. In producing this study, the author was fortunate to gain access to international actors at the center of many of the case studies presented.

The period of 1963–1992 in United Nations environmental politics was chosen as the primary empirical focus because of the availability of archival documentation to facilitate the detailed process-tracing

[70] Skocpol 1979, 380.

investigation undertaken. With many archives following a "30-year rule" for the release of documents, the period under study could be analyzed in detail. To increase the number of observations of variables under study, however, I have included a short treatment of the post-1992 period, covering events that included the 2002 World Summit on Sustainable Development and the 2012 "Rio+20" United Nations Conference on Sustainable Development. While different in character than the detailed, process-level analyses undertaken in Chapters 3–6, this approach enables further testing of the framework and is of empirical interest.

My analysis has, furthermore, focused – broadly – on UN environmental institutions. In this respect, my empirical focus is similar to what Kate O'Neill calls the "meta-regime" for Global Environmental Governance and Lars-Göran Engfeldt calls the "Stockholm-Rio-Johannesburg" process.[71] I broaden this focus somewhat in considering negotiations in the field of sustainable development, mainly to the extent that these discussions situated environmental cooperation in a wider social and economic context. I address important negotiations surrounding Multilateral Environmental Agreements, such as the UN Framework Convention on Climate Change and the Convention on Biological Diversity, in only a tangential fashion. I have adopted this focus largely as a means of rendering the scope of my empirical examination into more manageable proportions. As I have shown elsewhere, however, the theoretical approach taken in this book is also applicable to the pattern seen in the negotiation of environmental treaties and conventions.[72]

Conclusion

Existing institutional analyses can be complemented through a fuller examination of the part that temporal coordination problems play in institutional change processes. Leading theoretical treatments assume that coordination in the presence of sufficient change incentives is largely unproblematic. Yet, as demonstrated by the preceding analysis, combined with high transaction costs, the gradual accumulation of incentives to reform institutions creates significant coordination problems that must be taken into account in explanations of institutional

[71] Engfeldt 2009; O'Neill 2007. [72] Manulak 2020, 1–27.

change. This difficulty helps account for the widely observed status quo bias in institutional life. This chapter has outlined in detail the part that temporal coordination problems play in hindering efforts to recast institutional norms, rules, and procedures. There are good reasons to believe that, in addition to other well-documented factors, such as lock-in effects or path dependence, temporal coordination dilemmas are a chief obstacle to the realization of Pareto-improving institutional change.

This chapter has theorized the role of coordination dilemmas in generating a risk-dominant, status quo bias in international institutions. As a means of coordinating the investment of scarce political and analytical resources, as well as solving pernicious bargaining dilemmas, actors rely on prominent temporal landmarks, such as TFPs, around which to concert their expectations. By helping states to overcome assurance problems, convergent expectations contribute to an information-rich environment and provide a focus for negotiations. Consequently, some moments in time are more likely than others to see significant institutional transformations. The arrival of such junctures is often precipitated by TFPs, such as important anniversaries or crises, that catch the attention of all concerned. While convergent expectations cannot produce agreement if there is no incentive for it, they enable actors to capitalize on the presence of those incentives when they are available.

3 *The Stockholm Conference and Institutional Change*

The United Nations Conference on the Human Environment (UNCHE) marked a turning point in the history of global environmental governance. In June 1972, the representatives of 113 countries conferred in Stockholm, declaring "the urgent desire of the peoples of the whole world and the duty of all Governments ... to defend and improve the human environment for present and future generations." According to conference participants, cooperation in the environmental field would be pursued alongside the UN's two other fundamental goals: the promotion of peace and of socio-economic development.[1] The Stockholm declaration, action plan, and organizational recommendations were manifestations of a sweeping change in the global environmental order, thus setting patterns of cooperation that would persist for decades.

The Stockholm conference was a moment of significant change within United Nations environmental institutions. Key countries built a coalition to capture unrealized benefits from cooperation in the environmental field. This chapter investigates the document record and published literature to analyze the UNCHE negotiations. Exogenous shifts gave rise to incentives to revise existing modes of cooperation in the environmental field. Although countries stood to gain from recasting institutions, temporal coordination problems hindered cooperative efforts to realize joint gains through the 1960s and early 1970s. It was not until the emergence of a Temporal Focal Point (TFP), the 1972 Stockholm conference, that countries were able to realize beneficial change. The TFP enabled a temporal convergence of expectations among states, giving great power to the North/South cooperative formula sketched by development experts at the 1971 Founex seminar. I conclude the chapter with a brief

[1] Report of the United Nations Conference on the Human Environment 1972, 3–5.

consideration of the explanatory power of the theoretical framework presented in Chapter 2.

Exogenous Change and the Stockholm Conference

The rationale for the Stockholm conference was provided by the mounting environmental problems associated with industrial and technological development. Although major environmental challenges had been present since the Industrial Revolution (and, to a lesser extent, before), the scale of these problems accelerated greatly after the Second World War. Public concern about air and water pollution, contamination from pesticides and chemicals, and rapid exploitation of natural resources necessitated political action. Many of these problems had important transboundary implications, demanding bilateral, regional, and multilateral collaboration. Combined with technological development and global demographic change, these factors altered the underlying conditions upon which states interacted, giving rise to incentives to cooperate.

A brief discussion of the scale of these transnational problems, as seen in the 1960s and early 1970s, is illustrative.[2] Water pollution was a highly salient global environmental concern at the time. It was estimated that 2.1 million metric tons of oil was being released into the world's waters annually.[3] The prominent Norwegian ethnographer and adventurer, Thor Heyerdahl, brought the effects of ocean pollution to life anecdotally at the Stockholm conference with a well-publicized lecture.[4] On a regional basis, it was estimated that France and West Germany annually released 27 billion cubic meters of untreated liquid pollution.[5] Pesticides such as dichlorodiphenyltrichloroethane (DDT), spotlighted by Rachel Carson's *Silent Spring*, made their way into the air and waterways.[6] In one dramatic instance, 200 pounds of endosulfan – a suspected human teratogen – fell off a Rhine barge in 1969, causing scores of dead fish to wash up on Dutch shores.[7] As anxiety mounted concerning ozone depletion and acid precipitation, studies emerged pointing to the impact of increased concentrations of greenhouse gases on the global climate.

[2] For a detailed account of the contentious nature of environmental politics in this period focused on organizational questions, see: Ivanova 2007.

[3] *The Ecologist* 1972, 90. [4] Ward et al. 1981, 45–64.

[5] Rowland 1973, 114. [6] Carson 1962. [7] Ward and Dubos 1972, 69.

Concentrations of carbon dioxide (CO_2) in the atmosphere had increased by an estimated 0.2 percent per year since 1958, leading to an anticipated total increase of 18 percent by 2000. If CO_2 concentrations reached those projections, the mean surface temperature on the planet would increase by a projected 0.5°C.[8] In 1972, Barbara Ward and René Dubos observed: "virtually all scientists would agree" that if international action was not taken to curb global emissions, human industrial activity could have "profound and unforeseen effects on the earth's climate." The chief challenge, they contended, would be the negotiation of binding emissions reductions on a North-South basis.[9]

The development of sophisticated systems analysis methods also played an important part in the recognition of global environmental crisis. Dennis and Donella Meadows of the Massachusetts Institute of Technology modified Jay Forrester's World Dynamics model to make a dramatic case for restricting economic growth. In *Limits to Growth*, published just before the Stockholm conference, they contended that growth in population and industrial production would soon outstrip the earth's carrying capacity, forcing a sudden and uncontrollable decline in population and industrial activity.[10] Thus, industrialized countries should aim to limit growth and developing countries should perforce content themselves with semi-development. Their projections had an important effect on public and government opinion, in part because of the prestige of computer-generated quantitative methods at the time.[11] In addition to their prominent book, the MIT team circulated their findings informally to opinion leaders, including the *Ecologist* magazine, which authored *A Blueprint for Survival* and used many of the MIT group's findings to make their case. In spring 1972, Dennis Meadows and Jørgen Randers gained further exposure by publishing their research in, among other venues, *International Organization*.[12]

To cope with this growing international problem, states weaved an ad hoc web of institutional arrangements. Though the Stockholm conference is often seen as the starting point for UN environmental cooperation, this is inaccurate. The institutional landscape was, in fact,

[8] *The Ecologist* 1972, 82.
[9] Ward and Dubos, 1972, 195.
[10] Meadows et al. 1972.
[11] Strong 2001, 117.
[12] See: *The Ecologist* 1972; Meadows and Randers 1972, 213–236. Their findings also influenced: Choucri 1972.

already a fairly crowded place. Among other UN agencies, the World Health Organization (WHO), the United Nations Educational, Scientific and Cultural Organization (UNESCO), the International Marine Consultative Organization (IMCO), the International Atomic Energy Agency (IAEA), the Food and Agriculture Organization (FAO), and the World Meteorological Organization (WMO) all had burgeoning environmental mandates.[13] Outside of the UN, the Organisation for Economic Co-operation and Development (OECD) and the North Atlantic Treaty Organization (NATO) had also established a presence in the environmental field by the early 1970s.

This fragmentary order did not, however, correspond with the interconnectedness of planetary problems. One diplomat remembered there was "a growing number of scientists, environmentalists, responsible citizens and government officials who believed that many of the problems, often seen as single, self-contained issues, were, in fact, related and had far-reaching effects."[14] Yet, many international organizations – and government departments, at the national level – resisted the classification of the "environment" as a single sphere of institutional activity. Uncoordinated, fragmented systems of environmental cooperation were increasingly understood as incapable of addressing the challenges presented by industrial and technological development.

Progressive environmental deterioration inspired the Swedish government to propose an international conference on the human environment. In addition to being motivated by growing problems of transboundary pollution, such as acid rain, the Swedish government aimed to ease Cold War tensions by providing a basis for constructive East-West cooperation and by rejuvenating the UN.[15] The proposed conference, approved by the United Nations General Assembly in December 1968, sought to "provide a framework for comprehensive consideration" of environmental issues and to "focus the attention" of governments on areas where they might cooperate in this field.[16]

On the surface, the Stockholm initiative seems to fit neatly into an explanation about the role of exogenous transformations in

[13] For a theoretical analysis of how international organizations move beyond their initial mandates and respond to evolving international problems, see: Hall 2016.

[14] Bruce 1991. Geoffrey Bruce Interview with the author, November 15, 2012.

[15] Bernstein 2001, 139–140; Engfeldt 2009, 34.

[16] UNGAR 2398, 1968.

precipitating institutional change. Industrial and technological development presented a set of problems that the international community was unable to address through existing institutional arrangements. The emergence of problems, especially transboundary air and water pollution, gave rise to incentives for states to gather and design the necessary rules and procedures to better manage the global commons. The launching of the United Nations Environment Programme (UNEP) following the Stockholm conference seems to confirm this assessment.

A closer inspection of the Stockholm case belies this straightforward explanation. Mounting environmental change caused by industrialization *did* produce the global problems that ultimately necessitated the Stockholm conference. These problems emerged gradually, however. It was a slow-moving crisis. Though the effects of accumulating pollutants were becoming increasingly visible, there was no sudden shock or crossing of an ecological threshold that persuasively explains the timing of institutional change in 1972. Although environmental problems were probably worse in the early 1970s than they had been in the late 1950s or early 1960s, this was a matter of degree. The list of environmental problems facing states in the early 1960s did not differ markedly from those confronting states in the early 1970s.[17] The constellation of interests within UN environmental institutions did not change in the lead-up to the Stockholm conference. Thus, judged from developments in external conditions, there is no reason why the early 1970s stand out as the clear time to bring about change. Exogenous shifts, rather than shocks, produced the incentives to realize institutional change.

The year 1972 was not, furthermore, the first time that leading actors advocated international environmental cooperation. At his September 1963 address to the UN General Assembly, US President John F. Kennedy called for greater global cooperation on environmental issues. Kennedy's proposal included a global wildlife and forest

[17] The agenda for a UN environmental conference in the early 1960s may not have been significantly different than that of UNCHE. While scientific knowledge of environmental problems had increased significantly before UNCHE, much of this was actually stimulated by the conference process itself. Collaboration with the International Council of Scientific Unions, as well as the Study of Critical Environmental Problems and the Study of Man's Impact on Climate, helped produce many crucial working papers. These studies were initiated by the UNCHE secretariat.

conservation strategy, cooperation on combating global epidemics, and an initiative to promote farm productivity and food distribution.[18] The American statement to the UN General Assembly, in many ways, anticipated the common interest of developed and developing countries in world environmental governance through the United Nations. Through much of his term, Kennedy's successor, Lyndon B. Johnson, advanced efforts to curtail global population and spoke to the need to safeguard the international environment. In a February 1965 message to Congress, Johnson deplored the fact that humanity had "altered the composition of the atmosphere on a global scale through radioactive materials and a steady increase in carbon dioxide from the burning of fossil fuels."[19]

The Status Quo and Alternative Equilibria

Progressive changes in the international setting altered the cooperative basis for United Nations environmental affairs. Exogenous shifts in the decades preceding the 1970s had contributed to a growing world environmental crisis. Anthropogenic activities had a detrimental effect on earth systems and had given rise to the need for greater global cooperation. Institutional arrangements – a loose complex of programs and initiatives run by UN specialized agencies – were seen as incapable of addressing the scale and scope of planetary problems. The poor fit between existing institutional measures and the problem-setting created incentives for a transformation of behavioral norms, rules, and procedures. This section focuses on the interests of key actors and on the coordination dilemmas that they faced in realizing joint gains. Particular attention will be devoted to the risk-dominance of the prevailing equilibrium.

The Payoff-Dominant Equilibrium

Modern technology and industrialization had highlighted the interdependence of ecological systems and the finiteness of the biosphere.[20] In the decades that preceded UNCHE, the world community gradually became capable of upsetting the planetary equilibrium through the

[18] Kennedy 1963. [19] Johnson 1965. [20] Lee 1972, 339.

transnational impact of activities within their borders.[21] In particular, atmospheric and marine systems presented a set of problems that were truly global in scope. Seventy percent of the earth's oceans and atmosphere existed outside national jurisdictions.[22] Marine pollution issues were especially salient at Stockholm. As Ward and Dubos held, "it is, above all, at the sea edge that the pretentions of sovereignty cease and the fact of a shared biosphere begins, more strongly with each passing decade, to assert its inescapable reality."[23] It was the recognition of global interdependence and vulnerability that inspired the title of Ward and Dubos' report to the conference, *Only One Earth*, which functioned as the official slogan of UNCHE.[24]

Demographic and consumptive patterns fueled concern about resource scarcities. Highlighting the fragility of earth systems, the United States' UN Ambassador, Adlai Stevenson, spoke to the General Assembly in 1965 of tiny, interdependent, vulnerable "spaceship earth."[25] The population of the spaceship had nearly doubled between 1900 and 1960, from 1.61 billion to 3.02 billion. This led many scientists and demographic experts to advocate zero population growth policies.[26] Uncertainties about the reliability of global population data and about the earth's true carrying capacity made these problems seem even more ominous.[27]

Whether it was because of the interdependence of global systems or the limitations of the earth's carrying capacity, world environmental problems necessitated a genuinely *global* response. It was in the interest of states, North and South, to arrest planetary decline. This included engaging cautious developing countries in any policies that aimed to address environmental questions. Many of these countries had high rates of population growth and utilized antiquated, environmentally damaging technologies. Developed states also stewarded ecologically significant resources, such as the Amazon rainforest, or had large populations that could imperil earth systems.[28] Yet, no comprehensive means of addressing global environmental issues in a multilateral

[21] Strong 1972, 76. [22] Strong 1972, 77. [23] Ward and Dubos 1972, 203.

[24] The environmental organization Friends of the Earth also wished to name their 1972 book in a similar manner. After some controversy, the nongovernmental organization settled on: Friends of the Earth 1972.

[25] Stevenson 1965, 90. Stevenson's speech helped to inspire: Ward 1966.

[26] Strong 2001, 115. [27] Choucri 1972, 173–185.

[28] Herter and Binder 1993, 11.

forum existed prior to UNCHE. If the developing countries were not involved in an international response, there would be little capacity to confront global problems and a United Nations conference would have few virtues.[29]

While global problems seemed academic to many observers in developing countries, these states faced pressing environmental problems of a local and regional nature. Poor sanitation and human settlement challenges were widespread and required international attention. As a result, the conference secretariat, noted UNCHE Secretary-General Maurice Strong in an interview with the author, saw its "main position" as being "supportive of the developing country position." While such a stance was seen as "too venturesome" by some governments, it was accepted "by most."[30] The environmental causes of disease, poor housing, and food insecurity were of particular significance to developing countries.[31] Rapid, unplanned urbanization brought squalor and filth to many cities, thus badly taxing social infrastructure. Soil degradation and growing populations made it difficult for many developing countries to feed themselves.[32] UNCHE occurred at the height of the 1968–1973 Sahel drought, which resulted in widespread crop failure and livestock death in five African countries.[33] Developmental "mega-projects," such as dams and massive irrigation schemes, brought a deterioration in environmental and social conditions.[34] Development, it was increasingly recognized, could result in "ecological boomerangs," which brought "unwelcome and sometimes unexpected considerations" into focus.[35]

The local and regional problems facing developing countries were of broad international significance. Developing countries, furthermore, frequently lacked the financial and technical capacity to manage these issues effectively. The economic position of much of the developing world was unfavorable in the early 1970s. Many countries were heavily indebted and relied on primary products for the bulk of their exports.[36] International cooperation held the prospect of generating

[29] Kennett 1972, 34–35; Stone 1973, 100–101. For an authoritative general assessment, see: Gardner 1972.

[30] Maurice Strong Interview with the author, August 20, 2010.

[31] Lee 1972, 338. [32] Ward and Dubos 1972, 156.

[33] Clarke and Timberlake 1982, 39. [34] Ward and Dubos 1972, 168–169.

[35] Lee 1972, 339.

[36] Developing country debt had grown from US$20 billion in 1961 to US$50 billion in 1968. See: Ward and Dubos 1972, 171.

resources that could help these countries address their environmental problems. Securing technical assistance and technological transfer from industrialized states could greatly ease the human and ecological challenges facing poor countries.

Establishing a basis for a concerted global attack on planetary problems was also appealing for a number of non-environmental reasons. Because the possible proliferation of pollution standards of global significance had the potential to unsettle international economic relations, setting global rules and principles in this area would contribute to greater certainty in economic development.[37] Some governments, particularly in the global South, looked favorably upon the possibility of capitalizing on their excess "environmental assimilative capacity" to attract investment. Since many developing lands were relatively unpolluted, a wider disbursement of polluting industries might better utilize the earth's natural capacity to assimilate and neutralize waste residuals.[38] Strict regulation of polluting industries in wealthy societies could facilitate this process. Other observers noted the potential ecological benefits of using natural products, such as rubber, rather than synthetics.[39] If world markets turned increasingly to natural products, this prospect would bolster export markets for some developing world economies.

It was hoped that the commitment of developed countries to global environmental cooperation would also lead to an increase in development assistance allocations. As will be discussed in greater detail below, development assistance flows had ebbed in the late 1960s and early 1970s. If "Only One Earth," the slogan of the Stockholm conference, became a global ethic, many hoped that it would reinforce a commitment to poverty amelioration in the planet's poorest regions.[40] After all, many environmental perils of the developing world were linked intimately with human conditions. Wider and deeper collaboration through the UN could also bolster the world body at a time when it was in the throes of a credibility crisis.[41] Since the United Nations had become the principal global forum for promoting development, a rejuvenated world body could have important benefits for the developing countries.

[37] GATT 1972. [38] d'Arge and Kneese 1972, 423, 434.
[39] Rowland 1973, 73. [40] d'Arge and Kneese 1972, 423; Stone 1973, 106.
[41] Rowland 1973, 4.

Establishing an international basis for environmental cooperation held important potential benefits for developed and developing countries. The world environmental crisis and the clear inability of existing institutional arrangements to handle the problem gave rise to incentives for institutional change. Issues such as marine pollution, acid rain, world population growth, ozone depletion, and climate change became increasingly prominent. Local and regional environmental concerns also threatened world development. Institutional change in the UN environmental sphere, therefore, could yield a higher payoff than the status quo for all sides. Institutional alternatives were *payoff-dominant*.

The Existing, Risk-Dominant Equilibrium

Despite important draws, the response of the international community to environmental despoliation and institutional deficiency was slow to materialize. Change was hindered by severe temporal coordination problems, slowing states' realization of joint gains. This section highlights the important risks that industrialized and developing countries faced in backing change in United Nations environmental institutions. These risks contributed greatly to the persistence of the existing risk-dominant equilibrium.

Developing countries feared that international environmental regulations, if enacted in a manner that was insensitive to their interests, could retard development. Global environmental action, some feared, would freeze the current unequal global distribution of wealth.[42] Given gross inequalities and widespread poverty in developing lands, this was a serious concern. There were, these countries jabbed, two classes of passengers on spaceship earth.[43] Large sections of Africa, Asia, and Latin America were seeking the benefits of industrialization.[44] Environmental requirements could increase the costs of development, since the ecologically damaging shortcuts taken by wealthy societies

[42] de Araujo Castro 1972, 404. [43] de Araujo Castro 1972, 403.

[44] It was estimated at the time, somewhat arbitrarily, that when a country's annual per capita income surpassed $500, it would emerge from the pretechnological condition. Of the countries with an annual per capita income below that level, 80 percent of them had an annual per capita income below $250. See: Ward and Dubos 1972, 146.

might have to be foregone.[45] Some worried that talk of limits to growth was "only a ploy to tell the developing world that they should not start consuming resources which were badly needed for the whole of mankind and which were getting depleted at a fast rate."[46] Internal Brazilian preparatory documents, for instance, indicate that the principal aim of its influential diplomacy during the preparatory process was to promote unfettered economic development as a chief solution to environmental problems in the global South.[47]

Environmental concern came in a period of particular economic insecurity in many developing countries. Because of widespread disillusionment among donor states with the disappointing returns on past international development projects, aid flows had shrunk considerably. The "liberal hour" of generous funding, it was feared, had passed.[48] The developing countries, as Mahbub ul Haq recalls, feared that "the industrial countries' concern with environmental problems was an alibi for not meeting the Second Development Decade's targets."[49] This worry manifested itself in a desire to keep the UN agenda focused on development, so that the interests of the poor countries would not be "summarily brushed aside from the crowded agenda of the powerful and rich nations."[50]

Poor countries also feared that, without safeguards, environmental considerations could provide a pretext for nontariff trade barriers. Protectionist impulses might, for example, manifest themselves in policies in industrialized states that would aid domestic producers by restricting imports from countries whose environmental standards were lower. The agricultural sector was especially prone to this risk.[51] Fuel standards governing the sulfur content of oil could provide another point of tension between pollution abatement strategies and

[45] MacDonald 1972, 393. [46] ul Haq 1976, 80.

[47] Miguel Ozorio de Almeida, Memorandum para o Sr. Secretario Geral, "Meio Ambiente Humano. Conferência de Estocolmo. Evolução dos preparativos. Visita do Senhor Strong." January 18, 1972, Itamaraty, Government of Brazil.

[48] ul Haq 1976, 84. [49] ul Haq 1976, 123.

[50] A Canadian diplomatic cable summarizes this view: "the developing countries, rightly or wrongly, view the UN as 'their organization' and are not generally keen on it being 'diverted,' into new fields of endeavour not of primary interest to themselves. In this connection, they have the votes." See: PERMIS-NY to USSEA, Institutional Arrangements Flowing from Stockholm Conference on Environment," January 24, 1972, Library and Archives Canada, 68-4-UN-1972-7-CDA. See also: ul Haq 1976, 86.

[51] d'Arge and Kneese 1972, 423; GATT 1972.

current patterns of international trade. Oil from the Persian Gulf and Venezuela was, for instance, high in sulfur relative to Indonesian, North African, and Nigerian sources.[52] The rise of "green conditionality" in the provision of development assistance was another danger. The World Bank decision in 1970 to appoint an environmental adviser to evaluate the ecological impact of its projects caused alarm.[53] The potential intrusion of donor countries into national development plans would be unwelcome, even if it were supported by sound science.

The developed counties also faced important risks in realizing institutional change. Many governments in the global North feared that environmental regulations would increase the costs to home industries, damaging their international trading position, diminishing real national income, and eroding their long-term comparative advantage.[54] Without coordinate action by other economies on a wide international basis, some feared that "pollution havens" with lax environmental standards could attract dirty industries.[55] In addition to harming the economies of regulating states, this potentiality would negate the anticipated benefits of stringent environmental measures. The recognition of different – less onerous – environmental responsibilities for the developing world could, paradoxically, worsen global environmental conditions by effectively providing a license to pollute. If, moreover, governments were to establish an informational basis for global environmental assessment and management, broad international participation was a necessity.

Temporal coordination problems manifested themselves in the bargaining stance taken by leading states during the UNCHE preparatory process. Absent a focal time frame, these states adopted defensive negotiating positions despite a potential interest in environmental cooperation. Developing countries voiced particularly deep misgivings about UNCHE's aims. In spring 1970 at a UN Committee on Programme and Coordination meeting, Brazilian first secretary, Bernardo de Azevedo Brito, attacked the conference as a "rich man's show."[56] At the summer 1970 session of the UN Economic and Social Council (ECOSOC), the Brazilian representative took this criticism a step further, accusing the industrialized countries of deliberately

[52] MacDonald 1972, 388.
[53] de Araujo Castro 1972, 412–413. See also: Le Prestre 1989.
[54] d'Arge and Kneese 1972, 421–422. [55] Stone 1973, 115.
[56] Engfeldt 2009, 41.

distracting attention from the international development agenda and seeking to keep the poor countries in a perpetual state of economic stagnation.[57]

Brazil's ECOSOC statement cast a pall over preparations, causing deep concern among reform-minded actors. With the conference less than two years away and already considered behind schedule, its substantive basis had been called into question. In September 1970, the Canadian, Swedish, American, and British representatives met in New York to explore the possibility of coordinating their strategies for assuaging the concerns of the developing countries.[58] The main thrust of their approach was to emphasize the consistency of environmental considerations with the goals of the Second Development Decade and to underline the benefits to developing countries of not repeating the developmental mistakes of industrialized countries.[59] Developed countries believed that change in patterns of UN environmental cooperation would suit developing states' "real, although still largely unacknowledged, self-interest in joint action to improve the environment."[60]

The conference secretariat, led by UNCHE Secretary-General Maurice Strong, aimed to bridge the North/South divide. Strong scheduled an informal meeting of the preparatory committee from November 9 to 10, 1970, at which he addressed the "alleged conflict"

[57] C. O. Spencer Memorandum, "Human Environment: Views of the UN Secretariat," August 11, 1970, Library and Archives Canada, 68-4-UN-1972, vol. 5.

[58] CANDEL-NY to DEA-OUN and DEA-FSR, "UNGA XXV: Second Cttee: Human Environment (item 43): Brazilian Views," September 28, 1970, Library and Archives Canada, 68-4-UN-1972, vol. 5. For the most comprehensive analysis of Brazilian policy during the conference, see: Corrêa do Lago 2009.

[59] PERMIS-NY to DEA-FSR and DEA-OUN, "UNGA XXV: CDA/USA Mtg: Human Environment," September 15, 1970, Library and Archives Canada, 68-4-UN-1972, vol. 5.

[60] Memorandum from Acting Secretary of State Johnson to President Nixon, "U.S. Policy toward the Major International Organizations Dealing with the Environment," August 24, 1970, Foreign Relations of the United States, vol. E–1, Document on Global Issues, 1969–1972, Doc. 299; W. K. Wardroper Memorandum, "Commentary for the 25th United Nations General Assembly," August 13, 1970, Library and Archives Canada, 68-4-UN-1972, vol. 5; Bureau of Intelligence and Research (U.S.), "Major International Conferences on the Environment 1971–1972," October 8, 1970, Foreign Relations of the United States, vol. E–1, Document on Global Issues, 1969–1972, Doc. 300.

between environment and development.[61] In one of the most important interventions during the conference preparatory process, Strong associated environmental ends with human-centered modes of development. In doing so, he went further than leading states had so far been willing to go toward reconciling the positions of developed and developing countries. At the February 1971 preparatory committee meeting, the secretariat proposed the addition of "economic, financial and social aspects of environmental issues" to the conference action agenda.[62]

Strong's proposed "radical remake" of the conference agenda was, he recalls, greeted with equal parts openness and skepticism.[63] States were reluctant to surrender their defensive negotiating positions at this stage of the process. The risks were simply too high. Developed countries, which primarily sought to enable action on global environmental problems, continued to implore developing countries to learn from their past ecological mistakes. Dismissing these arguments, Brazilian Ambassador Joao Augusto de Araujo Castro exclaimed:

> We refuse to accept a purely "conservationist" approach which deals with peripheral problems without going into the roots of the evil. We refuse to accept a situation in which policies can be defined or even imposed outside the context of national plans and priorities for the economic and social development of developing countries. We refuse to accept suggestions, which do not take into account the responsibility of developed countries to provide additional economic and financial aid to developing countries to redress the imbalance provoked by past and present patterns of international trade ... We refuse to accept any kind of international measure in the field of environment which does not take into account the right of each country to freely exploit its own natural resources, which tries to impose and enforce solutions on matters which are of the sovereign competence of each country ...[64]

[61] Maurice Strong, "Opening Remarks by Maurice F. Strong, Secretary-General Designate, United Nations Conference on the Human Environment, at Informal Meeting of Preparatory Committee for the Conference," November 9, 1970, Library and Archives Canada, 68-4-UN-1972, vol. 6.

[62] Preparatory Committee for the United Nations Conference on the Human Environment, Second Session, February 8–19, 1971, A/CONF.48/PC.8.

[63] Strong 2001, 122.

[64] Statement by the Brazilian Representative, H. E. Ambassador Joao Augusto de Araujo Castro, on Item 43 of the Agenda – Human Environment," November 1970, Library and Archives Canada, 68-4-UN-1972, vol. 6.

Despite Strong's intervention, countries continued to resist the integration of developmental and environmental ends. Indeed, just months before the conference, Brazilian officials invited Strong to Rio to meet with the highest levels of their government to underline their threat to boycott the conference and to persuade other developing countries to do likewise.[65] Developed countries, for their part, held to their paternalistic line about environmental lessons and developing countries denied the relevance of UNCHE for their concerns.

As negotiations moved into 1971, "the Stockholm environmental debate showed signs of becoming a spirited confrontation pitting the North against its neighbours to the South."[66] At the root of the problem were coordination dilemmas. Influential developing countries refused to acknowledge the relevance of global environment and pollution issues to their interests. While, as noted above, developing countries *did* stand to benefit from global environmental cooperation, they faced significant risks in acknowledging their interests in reforming UN environmental institutions. Institutional change that failed to account sufficiently for the position of the developing countries was potentially worse for those countries than the status quo. Similarly, leading developed countries refused to validate the environmental agenda of the global South. Although international problems such as sanitation and desertification *were* relevant to developed countries, UN environmental institutions that, in essence, gave developing countries a free pass on environmental problems would be detrimental to the planet and to their economic standing. Coordination failure at UNCHE would freeze the status quo, waste scarce resources, and weaken cooperative players' bargaining position in future negotiations. If only one side recognized the validity of the other side's position, change efforts would fail and that side would badly compromise its bargaining position in future talks. This dynamic was at the root of the temporal coordination problem facing states prior to UNCHE.

[65] Miguel A. Ozorio de Almeida to the Secretàrio Geral Adjunto para Organismos Internacionais, Meio ambiente, Reunião dos 77 – Lima, November 5, 1971, Itamaraty, Government of Brazil.

[66] Herter and Binder 1993, 9.

The Stockholm Temporal Focal Point and Change

Although states, North and South, could realize joint gains through institutional change, efforts to alter current modes of cooperation were fraught with risks. Developed and developing countries refused to acknowledge the validity of the other side's concerns for fear of weakening their hand in future negotiations. Obstinate posturing, instead of purposeful negotiations, resulted. As UNCHE neared, however, the increasingly prominent Swedish conference emerged as a Temporal Focal Point. The urgent need for an agreement gave great power to the findings of the report of a development seminar, held in Founex, Switzerland, in June 1971. This section first demonstrates that UNCHE should be considered a Temporal Focal Point. I then proceed to explain the relationship between the focal time frame and the realization of change in UN environmental institutions in 1972.

The Stockholm Conference as a Temporal Focal Point

As noted in Chapter 2, Temporal Focal Points are defined by three main characteristics: they are discrete, unique, and conspicuous. In defining the Stockholm conference as a Temporal Focal Point, the first of these factors is uncontroversial. The Stockholm conference, scheduled from June 5 to 16, 1972, represented a discrete moment in time. It was clear to all actors when the UNCHE negotiating process had begun and would end. The four conference preparatory meetings and the production of official documentation were scheduled in relation to this timeline.

The Stockholm conference was also unique. There was no forum at the time that could compete with UNCHE in taking center stage for the advancement of global environmental cooperation. The disparate environmental activities of United Nations specialized agencies, such as FAO, WHO, or UNESCO, were on a relatively small scale and represented only a small part of those agencies' activities. Although the OECD and NATO aimed to develop environmental programming, these smaller fora could not form the basis for global cooperation. Within the environment sphere, the Stockholm conference would give environmental concerns unprecedented status. There was no competing forum that could at that time draw the eye of a wide array of international actors in a comparable manner.

The conference was also highly conspicuous. A striking transition occurred during the UNCHE preparatory process in the second half of 1969 and the first half of 1970. The conference shifted from an essentially technical gathering to become a highly political event. More fascinating still – with the partial exception of Sweden – this shift was propelled by factors largely independent of the principals to the negotiation. The United States took a very restrictive attitude toward the conference and its budget in the early stages.[67] The Soviet Union and many Western European countries were unenthusiastic about an ambitious United Nations conference.[68] In the early stages, developing countries were apathetic about a meeting that, in their view, was of limited relevance to them.[69] Although Sweden served as the de facto conference secretariat through 1968 and 1969, there is no evidence to suggest that Sweden aimed deliberately to engineer a landmark political event that was to unfold.[70] Sweden's efforts were a far cry from "hegemonic leadership."

While the conference secretariat was, as we shall see, a leading player through much of the Stockholm process, it did not enter the equation until autumn 1970. From March until November 1970, a Swiss engineer, Jean Mussard, led the UNCHE secretariat as "Director

67 In November 1969, for example, Russell Train, the Chair of the US Council on Environmental Quality, wrote: "signals have come to me from several sources that the United States is not particularly enthusiastic about the Conference, particularly reflected in our attitude toward the Conference budget." See: Russell Train (Council on Environmental Quality) to Elliot Richardson (Undersecretary of State), November 17, 1969, Foreign Relations of the United States. Even as late as December 1971, an internal Canadian memorandum noted:

> U.S.A. though publicly committed to the conference may have been inflexible on the [East German] participation issue since there were some in Washington D.C. who were not too enthusiastic about commitments made to a globally-oriented conference on the environment, preferring nationally-based or regional arrangements. U.K. privately was unenthusiastic partly owing to their doubts about any UN-sponsored conference and partly their preference at the present time for European-centered cooperation.

See: Geneva (Ignatieff) to DEA-ECS, December 21, 1971, "Stkhm Conf/Talk with Maurice Strong," Library and Archives Canada, 68-4-UN-1972, vol. 13.

68 Arculus to MacInnes (FCO UN), February 18, 1971, UK National Archives, FCO 55/833/SME2/522/1.Part 3.

69 PERMIS-NY to DEA-OUN and FSR, "UNGA XXV: CDA/USA MTG: Human Environment," September 15, 1970, Library and Archives Canada, 68-4-UN-1972, vol. 5.

70 Engfeldt 2009, 38.

of Studies." Mussard's prosaic title and his technical focus rapidly became unsuitable for the increasingly political conference. The speed at which actors' expectations changed, rendering anachronistic a mandate that had just been approved by the UN General Assembly months earlier, is notable. The decision in August 1970 to supplant Mussard with Maurice Strong, a Canadian development administrator with a keen political acumen, marked an important turning point in the UNCHE process.[71] This switch, recalled Strong, was driven by Swedish recognition that Mussard would be unable to manage the highly political event that was unfolding.[72] Although Strong would dramatically alter the tempo and tenor of preparations after he was appointed, he did not himself *cause* this initial shift in Stockholm's fundamental orientation.

The transformation of the Stockholm conference is explained in part by a series of occurrences that helped produce a temporal convergence of expectations among states.[73] Numerous dramatic events occurred either immediately before the Swedish General Assembly initiative or during the early stages of preparations, adding considerable impetus to UNCHE. The foundering of the *Torrey Canyon* supertanker on March 18, 1967, spilling 32 million gallons of crude oil into the English Channel, attracted international attention. The incident served as a dramatic warning about the dangers of modern technology. The *Torrey Canyon* had the capacity to transport 120,000 tons of oil. Just five years earlier, the average oil tanker had a capacity of 12,000–13,000 tons.[74] In 1968 Paul Ehrlich published *The Population Bomb*, which predicted mass starvation and social instability if governments did not take drastic action to curb world population.[75] In June 1969, the Cuyahoga River in Ohio caught fire owing to high concentrations of petroleum by-products in the water. *Time* magazine published a prominent article reporting the fire, and observing that the heavily polluted river "oozes rather than flows."[76] Many of these issues were highlighted when *Fortune* and *Time* magazines published special environmental features in February 1970.[77] Earth

[71] Herter and Binder 1993, 19.
[72] Maurice Strong Interview with the author, August 20, 2010.
[73] Much of this discussion draws on: Dr. Jim MacNeill Interview with the author, August 18, 2010.
[74] Ward and Dubos 1972, 200. [75] Ehrlich 1968. [76] *Time* 1969.
[77] See: Fortune 1970; Time 1970.

Day was first celebrated, with the backing of the UN Secretary-General U Thant, on April 22, 1970. The new public interest in environmental issues was further on display when, a few months later, the United States Army was greeted with an unexpectedly negative public reaction when it dumped canisters of lethal nerve gas into the Atlantic Ocean near the Bahamas.[78]

It is important to note that these demonstrative incidents did not reflect a sudden worsening of global environmental problems. Though oil spills, such as the Torrey Canyon disaster, had devastating effects on marine ecosystems, accidental spills accounted for less than 10 percent of the oil annually released into the world's waters.[79] The June 1969 Cuyahoga fire was neither the first nor the largest recent blaze on that Ohio river.[80] The US munitions dump in 1970 was at least the tenth such operation since 1964.[81] These events were thus only symptomatic of long-standing problems. They did, however, focus the attention of actors on environmental issues and the Stockholm conference. Internationally, the desire to act primarily manifested itself in a *general* sense of the importance of UNCHE, rather than in *particular* efforts to address the problems brought to light by these incidents.[82] Stockholm became an outlet for gradually mounting international environmental concern. The conspicuousness and visibility of UNCHE among states increased greatly and, as one observer noted, "within a year of the Swedish initiative by the General Assembly, it had become clear that events had to some extent overtaken planning for the conference."[83] The UN, another observer remembered, "would have to stimulate more than discussion in 1972 if it was to be taken seriously."[84] The Stockholm *phenomenon* was characterized by emergent properties irreducible to the sum of its parts.

In the months leading up to UNCHE, there was an explosion of political and analytical investments in evaluating the current set of cooperative structures and proposing alternatives. Important studies were conducted by the International Council of Scientific Unions (ICSU) and the International Union for Conservation of Nature (IUCN), helping to generate new scientific information – at the national and international levels – on worldwide environmental

[78] Lyons 1970. [79] *The Ecologist* 1972, 90. [80] *Time* 1969.
[81] Wagner 2004. [82] Stone 1973, 98. [83] Rowland 1973, 34.
[84] Stone 1973, 19.

impacts.[85] Academic analysts contributed to this evaluation, including Barry Commoner and Paul Ehrlich.[86] The Club of Rome held its first major meeting in 1970 and, as a means of influencing developments in Stockholm, published its first major report, *Limits to Growth*, in 1972. Outside of these contributions, governments conducted detailed assessments to gain ecological perspectives in time for the conference. This included consulting and requesting input from national environmental specialists. Economists and development experts began to weigh in on environmental questions. This burst of new data and discussion radically altered the information environment facing global actors as they gathered in Stockholm.

The conference was influenced also by rapidly shifting political dynamics surrounding environmental issues. Coalitions supportive of institutional change which, to that point, had remained largely latent, established links and mobilized to influence UNCHE. In addition to the production of new information and analysis, the activities of academics, scientific unions, conservationists, and popular media affected the political balance on environmental issues. Beyond these actors, nongovernmental organizations and interested political observers organized to impact the process. To take one example, Friends of the Earth was established in 1969 and, in 1971, became an international network. To stir public opinion at UNCHE, the organization published a heavily illustrated book as an "introduction to the politics of survival."[87] The growing salience of environmental questions was manifested most vividly perhaps in the massive nongovernmental "Environmental Forum" organized in Stockholm to coincide with UNCHE and the giant papier-mâché whale paraded through the streets of the Swedish capital. These popular dynamics spurred the US Congress to hold numerous hearings on environmental questions in the lead-up to the conference and led to the perception within the administration of US President Richard Nixon that the international environment was a vote winner.[88] Conference preparations also spurred the creation of bureaucratic units within governments that affected the political balance and national preference-clarification efforts.[89] In the United States, for example, an international

[85] For a discussion, see: Smith 1972. [86] Commoner 1971; Ehrlich 1968.
[87] Friends of the Earth 1992. [88] Hopgood 1998, 76.
[89] Herter and Binder 1993, 22.

environmental task force was established and the secretary of state created an Advisory Committee on international environmental questions.[90] In the lead-up to the Stockholm conference, India set up a National Committee on Environmental Planning and Co-ordination and Prime Minister Indira Gandhi increasingly associated herself politically with environmental issues at the national and international levels.[91]

Rather than producing convergent expectations, substantial political and analytical investments followed a temporal convergence of expectations. Convergent expectations, in effect, produced a *market* for intellectual, political, and norm entrepreneurship, motivating a wide array of governmental and nongovernmental actors to seek to influence institutional processes. As we shall see in the ensuing sections, many of these contributions altered the political bargain struck at Stockholm in fundamental ways. Rather than generating convergent expectations, many of these contributions were thus a product of a temporal convergence of expectations.

These developments help to explain how Stockholm rapidly transitioned from a technical conference to a political one in 1969 and 1970. The conspicuousness of the conference would also increase through the deliberate actions of the conference secretariat in 1971 and 1972. Once appointed, Maurice Strong aimed to increase greatly the profile of UNCHE. Just before being formally designated conference secretary-general, he insisted on and received a modest boost in the secretariat's budget and leveraged most of the funds earmarked for documentation – intended for a research-heavy technical conference – to fund expert consultations.[92] Describing Strong's role, the future secretary-general of the Brundtland Commission reflected in an interview: "The UN had never seen anything like him. He worked outside the norms of UN politics. He brought a whole new spirit to the thing. He reached out to the scientists, the NGOs, and changed the whole spirit of conference

90 Hopgood 1998, 71–74.

91 For a discussion of Gandhi's environmentalism, see: Rangarajan 2009, 299–312.

92 C. O. Spencer Memorandum, "Human Environment: Private Discussion with Mr. Strong," September 23, 1971, Library and Archives Canada, 68-4-UN-1972, vol. 5. In fact, Strong made his willingness to accept the position of secretary-general conditional on US backing of his budget demands. See: CANDEL-NY to DEA, "Environment: Informal Mtg of Prepcom and Strong's Plans," October 30, 1970, Library and Archives Canada, 68-4-UN-1972, vol. 5.

preparations."[93] Strong obtained a US$100,000 Dutch grant through the UN Office of Public Opinion to hire two public relations officers, who got to work boosting the visibility of the conference.[94] This type of public relations activity was unprecedented in UN circles. Strong was, as one interviewee stressed, "adept at using his position and the intense media attention. He realized that, if he stirred the pot a bit, he could speak to and influence publics directly."[95] Strong obtained additional funds from external sources for special projects, including scientific consultations and "parallel processes" undertaken outside the formal conference proceedings.[96] Through these efforts, Strong's secretariat helped to crystallize the Stockholm TFP.

The profile and prestige of the Stockholm conference was also augmented by the publication of *Only One Earth* just months before the conference. The book, authored by Barbara Ward and René Dubos, highlighted many major environmental challenges confronting states. The way Strong worked around the bureaucratic opposition of the UN Publications Committee to the book proposal is illustrative of his entrepreneurial efforts. In his memoirs, he writes:

> But of course I wasn't ready to give up. I asked a colleague with more experience in the UN if there was any reason why I, as secretary-general of the conference, could not commission a book to be written, as a report *to* the conference rather than an official UN publication. There seemed to be no explicit prohibition, though apparently it had not been done before. Accordingly, I set up a separate non-profit organization specifically to produce the book, raised money for it outside the UN and gave Barbara Ward and René Dubos the green light.[97]

Only One Earth became a best seller and was translated into twelve languages. The political importance of the book was not lost on governments, which quickly become leery of its potential impact and the influence of its message even before it was published.[98]

[93] Dr. Jim MacNeill Interview with author, August 18, 2010.
[94] Stone 1973, 39.
[95] David Runnalls Interview with the author, August 13, 2010.
[96] Herter and Binder 1973, 22, 50. [97] Strong 2001, 125–126.
[98] R. Arculus Memorandum, "Notes of Conversation with the Mr. Maurice Strong on 11 February 1971," undated, UK National Archives, SME2/522.1, Part B; R. Arculus, "Preparatory Cmte-Environment," undated, UK National Archives, SME2/522.1, Part B.

Strong traveled the world in 1971 and 1972, delivering speeches, building networks, and drawing attention to the Stockholm conference. As North-South tensions escalated during the preparatory process, Strong "bought something like an accordion of plane tickets and proceeded to travel the world as a salesman for the conference."[99] It was on one of these trips that he met with Indian Prime Minister Indira Gandhi, persuading her to attend UNCHE.[100] Gandhi's commitment to participate further raised the profile of the conference and, according to Strong, rendered a threatened developing country boycott of UNCHE "very hard."[101] Strong's January 1972 visit to Rio was, perhaps, his most successful. In a pivotal outcome of the conference, Strong obtained firm support from Brazilian President, Emílio Garrastazu Médici, for key conference objectives. All told, Strong visited more than thirty developing countries in the lead-up to Stockholm.[102] By encouraging environmental nongovernmental organizations to contribute to the preparatory process, he put pressure on governments to reach a substantive outcome.[103] The UNCHE secretary-general also gave a flurry of interviews and public presentations.[104]

The secretariat's efforts contributed to the conspicuousness of UNCHE. Stockholm brought unprecedented attention and exposure to global environmental issues, providing states a prominent window for action. In his memoir of the conference, Peter B. Stone, who served on Strong's media team, noted that this perception was fostered intentionally. The secretariat set out to create a highly conspicuous event, deliberately "making people feel that Stockholm was important and valuable," for:

> The more this feeling grew, the more Stockholm did become important and valuable. This is not to imply that it was any sort of confidence trick. It was a

99 David Runnalls Interview with the author, August 13, 2010.
100 Strong 2001, 127.
101 Maurice Strong Interview with the author, August 20, 2010.
102 Engfeldt 2009, 56. 103 Herter and Binder 1993, 35, 50.
104 Strong himself became something of a celebrity in the lead-up to the conference, as one member of his secretariat recalled: "I had not previously realized the extent to which Strong had become a honeypot to which all the bees around were flying. He was now the object of attention from all sorts of people who sensed that he was likely to become very influential indeed." See: Stone 1973, 127.

process perhaps comparable to the Law, in that the legal system only works if everyone – well, almost everyone – believes in it.[105]

Thus, the work of the secretariat helped crystallize a focal moment in time. The focal point became increasingly prominent as the June 1972 conference approached.

Thus, the Stockholm conference constituted a Temporal Focal Point in United Nations environmental politics. The conference, scheduled for June 1972, provided a discrete timeline for negotiations. Although UNCHE was initially intended as a technical, information-gathering affair, preparations were quickly overtaken by events. UNCHE became a political, action-oriented affair. A series of prominent developments in the lead-up to Stockholm, including environmental disasters, helped make the conference a conspicuous outlet for an international response to long-standing, gradually accumulating problems. When Maurice Strong assumed the role of UNCHE secretary-general, he deliberately aimed to create a sense of exceptionality about the conference: traveling the world, giving speeches, and building networks. All of these factors focused attention on the conference, making it a unique, conspicuous moment along the temporal continuum.

The Founex Report and Stockholm Outcomes

While the Stockholm conference had grown in prominence, negotiations were hampered greatly by the unwillingness of leading developed and developing countries to bargain cooperatively. Defensive posturing had dominated the UNCHE preparatory process and the decade preceding it, slowing progress in realizing accumulating incentives to alter institutional arrangements. If a viable North/South basis for global environmental governance was to be established under United Nations auspices, states had to overcome the risk-dominance of the existing equilibrium. Key to this process was the 1971 Founex seminar, which provided a bargaining framework for the resolution of the North/South impasse. After providing detail on the Founex meeting, I evaluate its importance for institutional change in 1972.

105 Stone 1973, 66–67.

As noted above, UNCHE Secretary-General Maurice Strong intervened in conference negotiations at a November 1970 informal preparatory committee meeting. Though his proposed broadening of the conference agenda had been well received, it "lacked the benefit of rigorous analysis and a well-developed policy thesis."[106] To flesh out his framework, Strong convened a meeting in New York in spring 1971 with a group of eminent development experts. Those present included Barbara Ward, Gamani Corea of Ceylon, Pakistan's Mahbub ul Haq, Saint Lucian economist Arthur Lewis, and Enrique Iglesias of Uruguay.[107] Because the meeting could not reach a definitive conclusion, the decision was taken to organize a seminar to examine the question in more detail.[108] Haq, Corea, and Iglesias worked with Barbara Ward and Marc Nerfin, Strong's chef de cabinet, to coordinate the meeting.

These events provided the backdrop for the June 4–12, 1971, meeting of the panel of experts on development and environment, convened in Founex, Switzerland, by the UNCHE secretary-general. Numerous prominent experts provided working papers that explored the relevance of environmental questions for development.[109] The working papers and seminar discussions covered a range of topics, including the impact of environmental considerations on international trade and development assistance flows, as well as the potential adverse consequences of environmental standards for less developed countries.[110]

The resulting Founex report argued that, unlike pollution problems seen in industrialized countries, the environmental challenges of developing countries were principally the result of underdevelopment.[111] Issues such as poor sanitation, disease, housing insecurity, inadequate nutrition, and soil erosion were rife in the developing world. The development process was central to resolving these issues. Founex reflected a growing disillusionment in the early 1970s among many development experts with traditional, growth-oriented economic

[106] Strong 2001, 123–124. [107] Strong 2001, 123–124.
[108] Corea 2008, 339.
[109] Eight of the seminar papers were published in: Founex Report (1972).
[110] For a detailed account of the Founex meeting, see: Manulak 2016.
[111] "Development and Environment: Report of the Panel of Experts Convened by the Secretary-General of the United Nations Conference on the Human Environment (Founex, Switzerland, June 4–12, 1971)" 1971.

strategies.[112] What was needed, the report stressed, was a "widening of the development concept" to include "urgent social and human problems." It was, moreover, in the interest of industrialized countries to provide resources to enable the developing countries to address these problems. In certain instances, environment-development trade-offs would have to be weighed by national governments. The report strongly reinforced the sovereign competence of countries to make these trade-offs and contained numerous references to the principle of "additionality." It suggested that donor countries ought to include additional development assistance funds to meet any environmentally motivated increases in the costs of development. If environmental standards entailed the movement of polluting industries to developing lands, this potentiality could be justified on a case-by-case basis. The report also largely steered clear of policy prescriptions for contentious population issues.

The Founex report served as the basis for a subsequent series of regional meetings organized in autumn 1971 by the UN regional commissions in Mexico City, Addis Ababa, Bangkok, and Beirut.[113] The meetings provided a means of ensuring that the developing countries read the Founex paper and also functioned as a "bellwether" for attitudes toward it among developing countries.[114] The Founex report was also added as an annex to UNCHE's official documentation, which was sent to participating governments. Maurice Strong did much to circulate the Founex findings. In the early part of 1972, he "went around to developing world leaders and stuck the report on

[112] A nice summary of this shift in thinking is provided by Salvatore Schiavo-Campo and Hans W. Singer:

> Previously, it had been taken for granted that expenditures on people, so-called social expenditures under such headings as health, housing, education, training, social services, etc., should be treated as consumption. They were considered the fruits of development, rather than its seeds. This carried the implication that countries serious about their development ought to reduce such "unproductive" expenditures in order to increase "productive" expenditures, building up the physical capital of the nation. Now social expenditures, previously labelled "unproductive," are viewed as equally important development expenditures, compared with economic expenditure.

See: Schiavo-Campo and Singer 1970, 70.

[113] "Memorandum: Panel of Experts on Development and Environment," March 31, 1971, Maurice F. Strong Collection, Harvard University Environmental Science and Public Policy Archives (ESPP), Box 37, Folder 371.

[114] Herter and Binder 1993, 25; Stone 1973, 108.

their desks. He would point to it and say 'these leading Third World economists said this.'"[115] In his meetings, Strong touted the document as a "historic turning point in environment development dialogue."[116] Because of his extensive travels, this method gave wide exposure to the report and its main ideas. In addition, as Steven Bernstein notes, the experts that participated in the June 1971 seminar "became political assets who helped convince developing world leaders to send delegates to Stockholm, and to attend themselves."[117]

The Founex report was well received by both developed and developing states. The divergent preferences of the parties, the paucity of mutually satisfactory solutions, and the urgency imposed by the fast-approaching conference all combined to give the Founex formula great resonance in North-South talks. Indeed, as the next section shows, this last aspect was especially important in the context of a TFP. Upon reading the report, Canada's permanent mission to the United Nations noted that it "strikes us as being extremely important since it may form the main basis [for negotiations]."[118] Similarly, Norway's permanent representative welcomed the report as providing a "manageable agenda" for addressing the development-environment issue.[119] Lead negotiators from the United Kingdom quietly regarded it as a "reasonable" document, with recommendations that were "much less contentious than most people had apparently expected."[120]

Brazilian Ambassador, Miguel Ozorio de Almeida, described the report's conclusions as "well-balanced, very objective and in general endorsed by Brazil." Bernardo de Azevedo Brito, Brazil's representative at the UN, argued that, by following the recommendations of these "world-renowned economists," countries could move beyond the "initial period of over-reaction" to environmental questions. Because of

115 David Runnalls Interview with the author, August 13, 2010.

116 CANEMB-MEX to DEA-ECS, "ECLA Environment Seminar Mexico," Library and Archives Canada, 68-4-UN-1972, vol. 11.

117 Bernstein 2001, 37.

118 PERMIS-NY to USSEA, "Stockholm Conference: Report on Development and Environment," August 17, 1971, Library and Archives Canada, 68-4-UN-1972, vol. 10.

119 "Statement by Ambassador Edvard Hambro, Permanent Representative of Norway to the United Nations," July 13, 1971, Library and Archives Canada, 68-4-UN-1972, vol. 11.

120 Kitching to Williams, September 27, 1971, UK National Archives, FCO55/837/SME2/522/1.

the "considerable movement" in the policy stance of Brazil, the Economic Commission for Latin America and the Caribbean endorsed the Founex group's conclusions.[121] At the September 1971 regional meeting of the Economic Commission for Africa in Addis Ababa, participating countries announced their unanimity in "congratulating the authors of the Founex Report on the insights that they brought into the environmental problems."[122] As one observer reflected, Founex was "regarded [by both developed and developing countries] as the political compromise necessary to bridge the development/environment question," forming a basis for institutional change.[123]

The Founex report became a critical basis for the Stockholm declaration and action plan. Although the policy angles, in most cases, did not originate at the Swiss seminar, the Founex report clustered together a set of policy compromises that were broadly supportable at UNCHE.[124] The Stockholm declaration and action plan drew heavily on the Founex report's recommendations. Founex's integrative thrust, grounded in socially motivated development approaches, would be captured in preambular point 4 of the Stockholm declaration, which stated that "In the developing countries most of the environmental problems are caused by under-development ... Therefore, the developing countries must direct their efforts to development, bearing in mind their priorities and the need to safeguard and improve the environment." A similar point of emphasis is contained in principle 8, which notes that "Economic and Social development is essential for ensuring a favorable living and working environment for man and for creating

[121] CANEMB-MEX to DEA-ECS, "Latin American Seminar on Problems of Human Environment," September 7, 1971, Library and Archives Canada, 68-4-UN-1972, vol. 11.

[122] United Nations Information Service 1971.

[123] David Runnalls Interview with the author, August 13, 2010.

[124] The difficulty of balancing the emphasis on developmental and environmental impulses was the major challenge in bargaining on the Stockholm declaration. A look at the UK response to a secretariat questionnaire on the declaration is illustrative: "The stress on the need to take account of the economic and social implications, particularly for the developing countries, of environmental action seems the wrong way round ... Thus, the order should be reversed – social and economic actions should take due account of environmental considerations and this should be sufficient for the draft declaration." (Central Pollution United, Department of the Environment (D. P. Walley), "Brief for the U.K. Delegation," January 20, 1971, UK National Archives, FCO 55/431, SME 2/522/2 Part B.)

conditions on earth that are necessary for the improvement of the quality of life." Founex was the crucial reference point for these formulations.[125]

Principle 9 of the declaration reflected the Founex report's emphasis on the relationship between environmental deficiencies, underdevelopment and vulnerability to natural disasters. Principle 9 also reinforced the importance of development assistance in meeting these perils.[126] Principle 11 captures Founex's warnings concerning the potential rise of "neo-protectionism" or other possible hindrances to development.[127] The concept of additionality, given further weight by the Founex findings, was included in principle 12.[128] Just as in the Founex report, the Stockholm declaration showed a profound confidence in the ability of development planners to manage the delicate environment–development interrelationship in specific contexts.[129] The origins of principle 23, which provided a potential basis for developing country exemptions from international environmental standards, can also be traced to Founex.[130]

The Stockholm action plan would also be heavily influenced by the Founex formula. The action plan contained 109 recommendations, which were divided into 3 main types. Environmental assessment involved information evaluation and review, research, monitoring and information exchange. Environmental management entailed goal setting and planning, international consultation and agreements. Supporting measures included education and training, organization, financing technical cooperation, and public information.[131] Recommendations 103(a & b), 107, and 109 were particularly important to specifying the nature of North–South environmental relations. Recommendation 103(a) mirrored principle 11 of the Stockholm declaration in its condemnation of environmentally motivated trade restrictions. The corollary, controversial recommendation 103(b), suggested that states establish appropriate compensatory means of dealing with the potential detrimental effects of environmental policies on

[125] Sohn 1973.
[126] This point corresponds to: Founex Report, paragraphs 3, 4, and 8.
[127] Founex Report, paragraph 54. For analysis, see: Sohn 1973, 469.
[128] Founex Report, paragraphs 28–33.
[129] This was especially so with principle 13. For an analysis, see: Engfeldt 2009, 64.
[130] See, for instance: Founex Report, paragraphs 20 and 33.
[131] This section derives from Stone 1973, 87.

developing world exports. Recommendations 107 and 109 concerned the provision of additional international financing. The former recommendation proposed a study of suitable mechanisms for financing international environmental action. Recommendation 109 argued that environmental measures should not affect the flow of development assistance and that industrialized countries should meet any additional costs associated with integrating environmental questions in development schemes.

The Founex report formed the basis for institutional change in Stockholm. It provided the kind of specificity that Maurice Strong's November 1970 proposal had lacked, working out policy angles relevant to the UNCHE agenda. While many of the ideas were not new, Founex packaged them into a framework that brought value to developed and developing countries. By bettering each side's status quo alternatives, it enabled agreement. The declaration and action plan, combined with new organizational arrangements, provided the institutional covenant upon which developed and developing countries would interact in the post-UNCHE period. The resulting formula affirmed the relevance of environmental questions for developing countries, but established important safeguards. Going forward, developing countries would play an integral part in UN environmental affairs. The dramatic shift in the UNCHE process and the sudden change in the bargaining positions of key developed and developing countries owed, as we shall see, to the resolution of temporal coordination dilemmas in the presence of a Temporal Focal Point.

Explaining Founex: Temporal Focal Points and Institutional Change

How do we account for the power of the Founex formula? The Founex report was a vital means of establishing a North/South basis for global environmental governance. At Stockholm, states realized an agreement that fell within a bargaining range that was created by gradual exogenous shifts in the decades before UNCHE. While boundedly rational actors did not initially recognize the opening, Founex highlighted an agreement possibility that fell within the Zone of Possible Agreement of relevant actors and enabled institutional change in 1972. This section will demonstrate that, without the Stockholm Temporal Focal Point, actors would likely have gone on without recognizing or realizing mutually beneficial agreement opportunities.

The first point to be made is that the Founex report derived its influence from the Stockholm TFP. Aside from the fact that the Swiss seminar was conceived, convened, and funded by the UNCHE secretariat, it is crucial to recognize that the meeting was more of a political exercise, than a technical one. Instead of tackling esoteric, theoretical, or conceptual points, the Founex report's analytical thrust targeted the most pressing bargaining questions facing the conference. The seminar took the negotiating positions of key states as its analytical focus. Its length and tone made it accessible to the diplomats and decision-makers that would attend UNCHE. The regional representativeness of its participants further illustrates its political orientation. Tellingly, the secretary-general of the Brundtland Commission – a report that, in many ways, covered the same ground as Founex fifteen years later – remarked: "I don't think it stands the test of time as a technical economic document. I know we didn't use it on the Brundtland Commission." Nevertheless, he maintained that it "was a crucial political document."[132]

The secretariat also sought to influence seminar discussions. It is worth noting that those oriented toward robust government intervention in the process of economic development were numerous and well-represented among seminar participants.[133] Many of these individuals had long-standing personal and professional relationships, and were sympathetic to Maurice Strong's viewpoint. The influence of those who were inclined toward so-called human-centered modes of development was augmented when Gamani Corea and Mahbub ul Haq were, respectively, elected chair and rapporteur of the seminar. In addition, as Peter Stone of the UNCHE secretariat later observed, Founex was "long on economists and short on ecologists."[134] Because of this, the seminar focused mainly on economic issues, such as trade and aid, that were relevant to the apprehensions of major developing countries.

[132] Dr. Jim MacNeill Interview with the author, August 18, 2010.

[133] It should be noted that in an interview for this project, Strong denied that the Founex outcome was in any way preordained. He said: "They were far too sophisticated to sign onto something that the secretariat wanted. They signed onto the process. The meeting was very argumentative and controversial." Maurice Strong Interview with the author, August 20, 2010.

[134] Stone 1973, 112.

The timeline for the completion of the Founex report was influenced greatly by the fast-approaching conference. Though the range of debate had been narrowed somewhat by the deliberate selection of participants, seminar discussions were still quite contentious. Notes from the meeting reflect an inconclusive debate.[135] Those who did not agree with the conclusions of social interventionists did not change their point of view based on the Founex seminar discussions. Wilfred Beckerman, for example, the most vocal proponent of growth-focused development at Founex, left the meeting before its conclusion and, despite having his name on the published report, did not read the report until it was in its final form.[136] There were, another attendee recalled wryly, "few epiphanies in Founex."[137]

Perhaps this should not be surprising. The development/environment nexus was, after all, a relatively new field of study and required the normal period of scholarly gestation and discussion. The trouble was that "Strong really wanted a result."[138] Founex would have little impact if it came either after the Stockholm conference or too soon before. Governments needed time to read the report if it was to bridge the North/South divide at UNCHE and provide a substantive basis for institutional change. "They hadn't really come to a proper consensus at the meeting," recalled David Runnalls, a participant in the seminar, "No amount of arm-twisting really bought a true consensus … Eventually they just gave it to Mahbub [ul Haq] and he took it to New York and produced the Founex report."[139] Timing was everything. Temporal-pacing heavily shaped deliberations.

The report's clarity of expression and argumentation came from the fact that its contents were only loosely based on actual seminar discussions. If it was not for the approaching conference that was quickly becoming a landmark political gathering, the seminar may well have been inconclusive. Despite having their names attached to it, seminar participants did not even see – let alone comment on – Haq's final draft

135 Maurice Strong Founex notes, June 5, 1971, Strong Papers, ESPP, Box 37, Folder 371.

136 Professor Wilfred Beckerman Interview with the author, June 11, 2011.

137 David Runnalls Interview with the author, August 13, 2010.

138 David Runnalls Interview with the author, August 13, 2010.

139 David Runnalls Interview with the author, August 13, 2010; ul Haq 1976, 81. This assessment was confirmed in: Lars-Göran Engfeldt Interview with the author, July 23, 2012.

until after the report had been widely circulated to governments and international organizations. In this way, the presence of a distinct, exceptional, time frame heavily influenced the production and content of the Founex report.

Founex became a political tool in efforts to forge a North/South coalition toward institutional change at Stockholm.[140] Once established as a basis for negotiations, offering value to all sides, the Founex compromise acquired a pull that made reformulation problematic, lest diplomats reopen messy, time-consuming bargaining. An intricate network of other arrangements, such as those surrounding post-UNCHE organizational measures, was weaved in a manner predicated on the Founex resolution to the North/South negotiation dilemma. Despite having significant sway among developing countries, Brazilian efforts at the third preparatory meeting to press developing country claims a bit beyond the terms of the Founex paper received "no support" from developed and other developing countries.[141] Peter Stone was incredulous that, at Stockholm, "diplomats eventually began to say things like 'I think we should stick closely to the text of the Founex report here' or 'We should not go beyond the position of the Founex report.'"[142] Both coalitions saw Founex as a suitable basis for policy coordination; tampering with its recommendations risked an impasse. No serious thought was given, for instance, to continuing deliberations in the months following the June 1972 conference to provide a more considered analysis. Developed and developing countries moved simultaneously to embrace Founex.

140 The literature on the Stockholm conference has been unequivocal in its assessment of Founex's impact. Christian A. Herter and Jill E. Binder note that "in some ways [Founex] was as important as anything that was subsequently produced at the conference itself." Martin W. Holdgate, Mohammed Kassas, and Gilbert F. White concluded that Founex "destroyed the false idea that [environment and development] were necessarily incompatible, and began to convince the representatives of the developing countries that environmental concerns were both more widespread and more relevant to their situation than they had appreciated." John McCormick argued that Strong's efforts, principally through the Founex seminar, broadened states' conception of the environment as an issue area so as to include problems that they would have previously categorized as developmental. See: Herter and Binder 1993, 25; Holdgate et al. 1982, 7; McCormick, 1991, 92–93.

141 Kitching to Mathieson and Williams, September 24, 1971, UK National Archives, FCO55/837/SME2/522/1.

142 Stone 1973, 102.

The impact of Founex was also increased by the fact that there were no other prominent negotiating formulas on the table to compete with it for attention as states examined their institutional options in the context of the Stockholm TFP. One of Strong's first acts as conference secretary-general was to reverse an earlier decision that UN specialized agencies would submit papers directly to the conference. He insisted that all official materials originate from the UNCHE secretariat; thus Strong's secretariat controlled the message reaching governments.[143] The secretariat devised an elaborate and, it must be said, opaque process for coding government national reports and basic papers submitted to UNCHE so as to provide a basis for conference documentation. The secretariat was, moreover, able to synthesize the papers submitted to it, thus producing official documentation that was a manageable 600 pages. This was quite a feat, since the secretariat received more than 20,000 pages of documentation – a ten-foot-tall pile of material.[144] All of this contributed to a high degree of coherence in the message reaching governments, adding to the strength of the Founex formula.

With a shrinking timeline and a widespread interest in taking action within the context of the Stockholm conference, governments embraced the Founex formula. Its influence came mainly from its usefulness rather than its causal or normative contribution to debates. There were significant limitations in the report's findings. As W. M. Adams writes, Founex was a "statement of faith," that development and environment "could be combined in some way which could optimize ecological and economic systems, without explaining how." The panel did not "break new conceptual ground."[145] In a similar vein, Steven Bernstein notes that, while Founex may have prevented UNCHE from becoming a political failure, it simply "juxtaposed" environment and development.[146]

The most important example of this juxtaposition was the fundamental and inescapable tension between the integrative message of the report – stressing the compatibility of environmental measures with human-centered development approaches – and the inclusion of the "additionality principle" within the report. The concept of additionality, one interviewee noted, "was in blatant contradiction with the rest

[143] Engfeldt 2009, 55. [144] Herter and Binder 1993, 29–30.
[145] Adams 2001, 56. [146] Bernstein 2001, 40.

of the Founex report."[147] Its inclusion reflected the political nature of the seminar. The presence of the additionality principle in the report was a "sop to the Brazilians," who were invited to send a diplomatic representative to the seminar as a means of bringing them onside with the secretariat's vision.[148] Certain ambiguities and contradictions were acceptable, therefore, if it meant that the report would have the desired effect on bargaining.

Founex's influence was derived from the need for states to establish some basis for North/South cooperation in the context of a Temporal Focal Point. No consideration was given to continuing talks after the conference so as to allow agreement to emerge slowly. As one participant recalled, "we had a high-level conference with high level political attendance, media, NGOs. Something had to be delivered ..."[149] States behaved *as if* there was a major temporal discontinuity after UNCHE that effectively took the "keep talking" option off the table. The coordination dilemma that had plagued negotiations in the decade preceding Stockholm was eased by the emergence of a prominent formula at the ideal moment. It was not so much the substantive content of Founex that mattered, but that UNCHE enabled a convergence of expectations *in time*. With the approaching conference, all sides were looking for a solution and knew that the other side was doing the same. Recognizing the availability of change incentives and a formula that could form the basis for agreement, both sides moved to realize institutional change in 1972. States were coordinating on two dimensions: space and time.

The Stockholm declaration and action plan established an agreed foundation for international cooperation on environmental questions in the United Nations context. The UNCHE principles formed a "soft" legal basis for cooperation and the action plan served as a guide for international environmental activities. Key norms, rules, and procedures adopted at Stockholm also manifested themselves in recommendations to create the United Nations Environment Programme. UNEP had three main components: a secretariat, Governing Council, and Environment Fund.[150] The balance of influence between these organizational elements reflected the institutional agreement reached at

[147] Dr. Jim MacNeill Interview with the author, August 18, 2010.
[148] David Runnalls Interview with the author, August 13, 2010.
[149] Lars-Göran Engfeldt Interview with the Author, July 23, 2012.
[150] United Nations General Assembly Resolution 2997 (XXVII) 1972.

UNCHE.[151] While donor countries would heavily influence the Environment Fund, developing states would have predominant representation on the intergovernmental Governing Council. UNEP's institutional design was a product of the broader agreement achieved in 1972, safeguarding the interests of developed and developing countries.

Conclusion

At the Stockholm conference, states realized a fundamental reordering of United Nations environmental affairs. The UNCHE declaration and action plan, as well as organizational arrangements, transformed the bases for UN environmental cooperation. Institutional change was made possible by the existence of available joint gains, created by gradual exogenous shifts that had taken place in the decades before UNCHE. Pressing global environmental problems, such as air and marine pollution, motivated international action.

While gradual shifts in global conditions provided a basis for change, actors were slow to realize new cooperative potentialities. The risk-dominance of the prevailing equilibrium provides a compelling explanation for the slow rate of reform. Because actors feared compromising their bargaining position in future negotiations, posturing and defensive bargaining tactics prevailed. The turning point was the emergence of a Temporal Focal Point, the 1972 Stockholm conference, which eased coordination problems. A wide array of actors, including academic and nongovernmental players, made substantial political and analytical investments targeting influence at UNCHE. The fast-approaching conference brought a temporal convergence of expectations, allowing all actors to focus on a shared timeline for talks. The knowledge that the other side was looking for an acceptable solution at the same time reduced the risk of moving toward a new cooperative equilibrium. It was this shared search for a viable solution that gave immense power to the Founex report. The Founex formula's influence arose primarily from its usefulness in the presence of a Temporal Focal Point, rather than its substantive content.

The massive burst in environmental analysis and political activities centered on the Stockholm TFP were not the design of any one actor.

[151] For an analysis of institutional outcomes, see: Ivanova 2021.

The contributions of all actors were both generative of, and a response to, the wider, emergent social dynamics at play. Even the interventions of the conference secretariat, so important to the crystallization of the TFP, were a response to the powerful undercurrents transforming the planned technical gathering into a landmark political event.[152] The Stockholm TFP thus generated a temporal convergence of expectations, a collective phenomenon that, not unlike market forces, can be best understood when stepping back to view the wider picture on aggregate.

Empirical evidence from the period leading up to the 1972 Stockholm conference provides strong support for the theoretical propositions contained in Chapter 2. The arrival of a Temporal Focal Point triggered change in the presence of preexisting incentives to alter institutional arrangements. In addition to the clear power of Temporal Focal Points to focus the minds of actors, the chapter demonstrates how slow states were in recognizing the availability of joint gains. There was no change in the problem-setting or in the incentive structure facing actors that persuasively explains why UN environmental institititions changed in 1972 rather than, for instance, in 1963 when President Kennedy advocated global environmental cooperation. The chapter also sets up the institutional bases for cooperation in the UN environmental sphere as they came out of the Swedish conference. Chapter 4 will examine the first decade of the United Nations Environment Programme, the chief organizational legacy of UNCHE.

[152] In his interview with the author, for example, Strong described how his recognition of the fast-growing political importance of the environmental issue drove his interest in the conference and opened a pathway to his more overt interventions in the political process. Maurice Strong Interview with the author, August 20, 2010.

4 *UNEP and the 1982 Nairobi Conference*

The United Nations Conference on the Human Environment (UNCHE) forged a North-South bargain that enabled institutional change in 1972. By finding common cause, developed and developing countries built a coalition in support of the Stockholm conference declaration and action plan. UNCHE also produced a set of recommendations that led to the creation of the United Nations Environment Programme (UNEP) by the UN General Assembly on December 15, 1972.[1] UNEP was mandated to serve as the principal basis for implementing the conference action plan and for fostering conformity with its principles. In its first decade, UNEP faced formidable obstacles to fulfilling its mandate. Although the discordant international political context of the mid-to-late 1970s contributed much to the programme's difficulties, important limitations in the design of the new body hindered its ability to effectively manage the global environment.

As UNEP's first decade drew to a close, a UNEP Governing Council "Session of a Special Character" (UNSSC) was convened from May 10 to 18, 1982 in Nairobi, Kenya, to commemorate the tenth anniversary of the Stockholm conference.[2] The session reviewed progress on the implementation of the UNCHE action plan and made recommendations for the next ten years of institutional activity. At the Nairobi conference, leading African countries sought institutional change to address their environmental concerns and to remedy the organizational weaknesses of UNEP. The Kenyan conference constituted a Temporal Focal Point, contributing to a period of focus on the performance of the UN's chief environmental organ. Despite UNEP's well-established limitations, key cooperative parameters were largely constant between 1972 and 1982. As a result, the Nairobi conference did not lead to institutional change.

[1] UNGAR 2997 (XXVII), 1972.
[2] UNGAR 35/74 1980; UNGAR 36/189 1981.

This chapter first provides background on the organizational arrangements that came out of UNCHE, highlighting the distributive conflicts that shaped UNEP. After establishing the basic features of post-Stockholm institutional measures, I highlight the major challenges that the programme faced in global environmental management and coordination within the UN system. I detail negotiations for institutional change at the 1982 session, which served as a Temporal Focal Point. An analysis of the positions and underlying interests of leading actors explains the absence of a North-South bargaining range that might have otherwise enabled agreement at the Nairobi conference. I conclude the chapter with an assessment of theoretical implications.

The Institutional Design of the UN Environment Programme

In proposing the Stockholm conference, Swedish officials argued that no new organizational machinery would be necessary.[3] As preparations proceeded, however, views on this issue began to change. UN Secretary-General U Thant made a speech in Austin, Texas, in May 1970, advocating the creation of a UN environmental body.[4] Through 1971, the UNCHE secretariat planned and contributed to five informal workshops and seminars for diplomats, politicians, academics, and journalists to examine organizational options.[5] Those informal meetings shaped the detailed negotiations that occurred in 1972.[6]

Negotiations concerning the design of UNEP were influenced by the need to bridge two main interest cleavages: (i) the trenchant opposition of UN specialized agencies to the proposed creation of a dedicated UN environmental body, and; (ii) the developed-developing world split. Proponents of change had become frustrated with the existing, fragmentary operation of UN environmental institutions. As noted in Chapter 3, the UN environmental field was already a crowded place in 1972. An internal Canadian memorandum is illustrative of the frustration with the preexisting setup:

3 Engfeldt 2009, 68.

4 United Nations Archives and Record Management Section, UN Secretary General – U Thant Fonds, Box S-0883-0025, Folder 8.

5 In 1971, meetings were convened in Siniai, Geneva, Colorado, New York, and Sussex. For a short description of these meetings from one of the participants, see: Engfeldt 2009, 71–72.

6 Richard N. Gardner Interview with the author, September 22, 2012.

The specialized agencies are no longer organized to cope with today's set of international problems much less those that will be appearing tomorrow ... The specialized agencies are quarrelling among themselves, each trying to carve out a piece of the environmental problem for their exclusive jurisdiction and practically all of their energy is going to fighting one another rather than getting on with the job.[7]

Given the heavy institutional machinery and the paucity of results in the environmental field, the system of overlap, duplication, and turf battles was seen by many as untenable. Sweden's permanent representative to the United Nations, Ambassador Olof Rydbeck, told the UNCHE preparatory committee that conference planning had "confirmed the need for a central coordinating mechanism in the United Nations to provide political and conceptual leadership."[8] In his February 1972 special message to the United States Congress, President Richard Nixon identified a need to create "a centralized coordination point" for United Nations activities in the environment field.[9] For the United States, there were "genuine advantages in international cooperation [in environmental protection], and in the coming years there will be rapid growth in the need for such cooperation."[10]

UN specialized agencies aimed to protect their bureaucratic turf by undermining efforts to launch UNEP. Britain, France, Italy, and the Soviet Union backed this position, seeking to reinforce the standing of agencies where they had strong influence.[11] The autonomous agencies

[7] Dr. P. D. McTaggart Memorandum, "International Co-operation and Environmental Improvement: The UN Specialized Agencies," July 21, 1971, Library and Archives Canada, 68-4-UN-1972-7-CDA, vol. 1.

[8] "Statement by Ambassador Olof Rydbeck in the Preparatory Committee for the United Nations Conference on the Human Environment at Its Fourth Session in New York on Monday, 6 March 1972."

[9] United States Department of State, "President Proposes Creation of UN Voluntary Fund on the Environment," February 8, 1972, Library and Archives Canada, 68-4-UN-1972-7-CDA, vol. 2.

[10] "Acting Deputy Director of the Office of Environmental Affairs (Salmon) to the Secretary to the Council on Environmental Quality (Gibbons)," July 14, 1970, Foreign Relations of the United States, vol. E-1, Doc. 298.

[11] Other countries, particularly the Soviet Union, the United Kingdom, France, and Italy took a dim view of efforts to create a new UN environmental organ. In February 1971, British representatives sought to obviate discussion of organizational matters through explicit preparatory committee opposition. This position received considerable support from the Soviet Union, which had taken a restrictive position on the potential for a new UN environmental agency. In opposing new organizational machinery, British and French officials sought to

worked aggressively at the Stockholm conference preparatory meetings to discourage states from creating new and, in their view, unnecessary UN environmental machinery. At informal consultations on institutional questions held in Sussex, UK, for example, UN agencies prowled the corridors, seeking to persuade participants of the benefits of current, decentralized patterns of environmental cooperation.[12] Days before the Stockholm conference convened in June 1972, the Director General of the International Atomic Energy Agency (IAEA) circulated a series of unsolicited amendments to the organizational proposals that were under consideration at Stockholm. By reinforcing and highlighting the autonomy of the specialized agencies in the environmental field, the IAEA amendments would have weakened greatly the new programme's ability to coordinate UN environmental activities.[13]

Specialized agencies were also busy in Stockholm, seeking to vitiate institutional proposals. Their efforts were backed by Peru and India. After the conference, Christian Herter Jr. of the US delegation reported:

> All of the major Specialized Agencies spent a considerable amount of time, energy and effort in lobbying the developing world and certain Developed Countries (eg. UNESCO with Belgium, IAEA with the United States) against the institutional arrangements resolution. They were all fearful that the resolution would place constraints on their own jurisdiction in the environmental field and were responsible for the introduction of various crippling amendments. The United States Delegation and others were aware of these

safeguard the position of the International Marine Consultative Organization, the United Nations Educational, Scientific, and Cultural Organization, and the Organisation for Economic Co-operation and Development. See: Arculus to MacInnes (FCO UN), February 18, 1971, UK National Archives, FCO55/833/SME2/522/1.Part 3; GENEV to DEA-UNS, February 29, 1972, Library and Archives Canada, 68-4-UN-1972, vol. 14. See also: "Note by the United Kingdom Delegation: Institutional Arrangements," Peter S. Thacher II Collection, Harvard University Environmental Science and Public Policy Archives, Box 14, Folder 121.

12 Engfeldt 2009, 72.

13 IAEA officials visited embassies in Vienna, seeking to gain support for their approach. See: Vienna to DEA-ECS, June 13, 1972, Library and Archives Canada, 68-4-UN-1972, vol. 5.

lobbying efforts and in the vast majority of cases were able to prevent these amendments from passing.[14]

Perhaps more important than their lobbying at Stockholm, agencies could resist coordination by the new environmental program going forward. Though states formally controlled UN bodies, it was difficult for them to do so unless they were able to harmonize their instructions to national representatives across numerous intergovernmental fora. Reining in these agencies through national-level coordination was a tall order, given the vigorous bureaucratic competition endemic to national governments.[15] The specialized agencies thus had an important influence over the conference's institutional provisions because they could render them ineffective in the longer run.

The need to gain the cooperation of obstructive UN specialized agencies narrowed the menu of organizational alternatives that were considered at Stockholm. In many ways, the resulting structure represented the path of least institutional resistance. To reassure the specialized agencies and account for the broad integrative nature of environmental questions, the proposed organization would serve a policy review and coordination role. UNEP would become a catalytic body, conducting environmental assessment and stimulating the work of other UN bodies. It would be a programme, not an autonomous specialized agency, and would not have operational functions itself. A small programme focused on facilitating greater coordination within the United Nations system was seen by the advocates of UNEP's design as an antidote to the heavy, inefficient machinery of the UN at the time.[16] To further allay the fears of the specialized agencies, the UNCHE secretary-general, Maurice Strong, was compelled to make a statement in Stockholm stressing that UNEP would not have the authority to make direct recommendations to other UN agencies. In

[14] Christian A. Herter to Secretary of State Rogers, "Report on the United Nations Conference on the Human Environment," July 28, 1972, Foreign Relations of the United States, vol. E-1, Doc. 325.

[15] Strong noted the importance of this dimension in a May 1971 speech at the University of Toronto. He argued that "effective coordination within the United Nations system as now constituted, must involve coordination within national governments." "Speech by Maurice F. Strong at the University of Toronto," May 27, 1971, Library and Archives Canada, 68-4-UN-1972, vol. 4.

[16] As has been demonstrated by Maria Ivanova, UNEP's limited mandate was not seen as a means of limiting its reach or effectiveness. See: Ivanova 2010.

all cases, Strong underlined, agencies received instructions exclusively from their governing bodies.[17]

North-South bargaining – the second major cleavage in organizational negotiations – centered on three main questions: the location of the new body within the UN system, the balance of influence between the environmental secretariat and the new intergovernmental body, and the physical siting of the programme's headquarters. Debate centered on a US proposal, tabled in February 1972, which suggested the creation of an environment fund, secretariat, and environmental coordination board.[18] US officials argued that the fund should be managed by an environmental administrator, supported by a small secretariat that would be responsible for funding all new environmental projects in the UN system. If the environmental fund financed *all* new UN environmental activities, it would exercise greater control over the environmental programs of specialized agencies.[19] The secretariat's administrative and program budget would be supported by the voluntary environmental fund. Further oversight and coordination could be performed by a twenty-seven-member intergovernmental functional commission of the UN Economic and Social Council (ECOSOC) and an interagency Environmental Coordination Board.[20] The environmental administrator should chair the latter body.

By maximizing the role of the environmental fund and the secretariat, donor countries would have a predominant influence over UN environmental activities. Having the secretariat's functions financed through the fund allowed donors to exercise leverage over it.[21] This financial structure also had the virtue of shielding the new environmental organ from the vicissitudes of the regular UN budget, which

17 "Final Report of the Canadian Delegation to the United Nations Conference on the Human Environment," Library and Archives Canada, 68-4-UN-1972-CDA, vol. 5.

18 For a discussion, see: Ivanova 2021, 25–54.

19 "Final Report on the Stockholm Conference," Library and Archives Canada, 68-4-UN-1972-CDA, vol. 5.

20 Given ECOSOC's mandate to support and facilitate the economic and social activities of the UN system, policy review and coordination in the environmental field was a natural fit for the council's work. US-led efforts in 1971 to renew and reform ECOSOC added further impetus to the reformer's goals.

21 Manulak 2020.

could hamper its activities.[22] In placing the intergovernmental body within the framework of ECOSOC, a body where developing countries had slightly less dominant representation, developed countries could exercise a greater degree of control than they might have if the body was constituted directly under the auspices of the UN General Assembly. Limiting the size of the intergovernmental body to twenty-seven and minimizing its power relative to the secretariat would increase the sway of industrialized countries.

Developing countries rejected the US proposal. A Brazilian counter-proposal aimed to ensure maximum control of environmental activities and the environmental fund by an intergovernmental body of the UN General Assembly.[23] An ECOSOC commission would be an inappropriate platform for the new organization.[24] Instead of the twenty-seven-member commission proposed by the United States, Brazil argued for a geographically representative membership of fifty-four.[25] The head of the environmental secretariat, Brazil maintained, should be a UN undersecretary-general, subject to the decisions of the intergovernmental body and fully integrated into the general UN secretariat. To ensure its independence from donors, it should be funded by the regular UN budget, rather than through the voluntary fund. While the fund could be used to coordinate UN agencies, a portion of it would also be used to fund any additional costs that developing countries incurred in tackling environmental problems. In their proposal, Brazil enjoyed the backing of leading developing countries.

To resolve the impasse, a drafting group, comprised of Brazil, Canada, Indonesia, Kenya, Sweden, and the United States, met to hammer out the set of recommendations that eventually gained unanimous support at UNCHE.[26] A small secretariat would be created,

22 "Instructions for the Canadian Delegation to the 4th Preparatory Committee for the United Nations Conference on the Human Environment," March 2, 1972, Library and Archives Canada, 68-4-UN-1972-CDA, vol. 3.

23 Relatório da Delegação do Brasil: À Conferência das Nações UNIDAS Sombre o Meio Ambiente, 1972, Itamaraty, Government of Brazil, 48–50.

24 "Statement by Ambassador Sergio Armando Frazão to the Second Committee," October 1971, Library and Archives Canada, 68-4-UN-1972, vol. 12.

25 PERMIS-NY to DEA-ECS, "Env: Post Stkhm Institutions and Fund: Brazilian/ Yugoslav Draft Resoln," April 21, 1972, Library and Archives Canada, 68-4-UN-1972-CDA, vol. 3.

26 "Final Report of the Canadian Delegation to the United Nations Conference on the Human Environment," Library and Archives Canada, 68-4-UN-1972-CDA, vol. 5.

headed by an executive director who was to be elected by the General Assembly upon the nomination of the UN Secretary-General.[27] While administrative expenses were to be met by the regular UN budget, programme costs, such as financing activities of specialized agencies, would be supported by the fund.[28] The secretariat was mandated to advise the intergovernmental body on the formulation and implementation of programs, while continually reviewing UN environmental operations. On the divisive question of the control of the fund, the environmental secretariat was mandated "to administer, under the authority and policy guidance of the Governing Council, the environmental fund."[29]

An intergovernmental committee, called the UNEP Governing Council, was established to oversee the activities of the secretariat and approve annually the utilization of the fund. The Governing Council would report to the General Assembly through ECOSOC. Though the drafting group had agreed on a Governing Council of forty-eight countries, a late Australian amendment in the conference plenary boosted this to fifty-four.[30] In December 1972, led by developing countries, the UN General Assembly further increased this number to fifty-eight. While the Federal Republic of Germany proposed criteria for election to the intergovernmental body, such as the size of contributions to the Environment Fund or demonstrated commitment to environmental cooperation, this provision was opposed by developing countries and defeated.[31]

[27] The election of the executive director was, at the time, an unusual procedure that aimed to give the position an independent political standing relative to the specialized agencies.

[28] The distinction between "hard core" and other administrative costs was a compromise advanced by the Canadian delegation to the conference. This was meant to leave the door slightly open for support of some types of administrative costs through the fund.

[29] "Final Report of the Canadian Delegation to the United Nations Conference on the Human Environment," Library and Archives Canada, 68-4-UN-1972-CDA, vol. 5.

[30] Christian A. Herter to Secretary of State Rogers, "Report on the United Nations Conference on the Human Environment," July 28, 1972, Foreign Relations of the United States, vol. E-1, Doc. 325.

[31] It appears that one reason that the United States came to abandon its position that UNEP should be a creature of ECOSOC was that, by associating the UNEP Governing Council with the General Assembly, West Germany could gain representation on the Governing Council. See: GENEV to USSEA, August 28, 1972, Library and Archives Canada, 68-4-UN-1972-CDA, vol. 5.

To advance coordination of international environmental activities, it was decided that the environmental fund should "finance wholly or partly the costs of the new Environmental initiatives undertaken within the United Nations system."[32] While modified from the initial US proposal that stipulated that the fund would finance *all* new UN environmental activities, the Stockholm recommendations provided a clear basis for UNEP to play a catalytic part in policy review and guidance. The proposed Environmental Coordination Board was established under the auspices of the UN's Administrative Committee on Co-ordination.

Developing countries argued that "additional" environmental costs should be met through the environment fund. This position was opposed vigorously by the industrialized countries, which argued that turning the fund into another avenue for development assistance would discourage contributions to it. Development assistance should, they maintained, be provided through existing channels.[33] On this issue, the developed countries held firm and gained acceptance for their position. Though "due account should be taken of the special needs of the developing countries," the fund supported environmental programs of general interest, such as monitoring, data collection and assessment, environmental quality management, research, and information exchange.[34]

Although the physical location of the new UN environment programme's headquarters was discussed at UNCHE, the issue was set aside until the General Assembly session. In December 1972, when the Stockholm recommendations were considered, the General Assembly selected Nairobi as the location of UNEP. While developed countries backed Geneva, based on the belief that a coordinating programme should be in close proximity to the agencies it was mandated to coordinate, Nairobi came to enjoy the unified backing of developing

[32] "Final Report of the Canadian Delegation to the United Nations Conference on the Human Environment," Library and Archives Canada, 68-4-UN-1972-CDA, vol. 5.

[33] "Instructions for the Canadian Delegation to the Informal Consultations to Be Held in Geneva May 3–4 on the International Organizational Implications of Recommendations for Action, Including Financial Implications of the United Nations Conference on the Human Environment," April 27, 1972, Library and Archives Canada, 68-4-UN-1972-CDA, vol. 3.

[34] UNGAR 2997 1972.

countries.[35] This support was based primarily on the political goal of achieving developing country representation among the administrative locations of UN bodies. Developed countries, furthermore, extracted a commitment from developing countries to support the election of an executive director from an industrialized country if the programme was sited in the developing world.[36] Maurice Strong, a Canadian, was duly elected the first Executive Director of UNEP.

The structure of UNEP was, therefore, a compromise between the goals of developed and developing countries. For developed states, UNEP enhanced their ability to coordinate the disparate activities of UN agencies, providing a mechanism to further international environmental cooperation. For developing countries, UNEP's structure offered a considerable degree of control over environmental activities. Indeed, as the internal Brazilian report on the conference observed, the strength of UNEP's intergovernmental body, including its control over the Environment Fund, was almost completely consistent with Brazil's aims on post-Stockholm institutional arrangements.[37] UNEP's large Governing Council and its location in Nairobi guaranteed that the priorities of developing countries would always be high on UNEP's agenda. Although the organizational structure of UNEP broadly resembled the US proposal of February 1972, industrialized countries made the bulk of the concessions on crucial details, which, as we shall see, were immensely important. What resulted was an organization that could be characterized as a "dual key" body, requiring support from developed and developing states to function effectively. The buy-in of specialized agencies was also vital.

UNEP's First Decade

As states approached the Nairobi conference, numerous problems emerged that gained a prominent place on UNEP's agenda. This section demonstrates that, when states reviewed the status of UN

[35] New Delhi also enjoyed considerable support, but India withdrew its candidacy after Kenyan President, Jomo Kenyatta, threatened to forcibly expel the many South Asians living in Kenya if India did not withdraw its candidacy. See: McDonald 1989, 9.

[36] Bruce 1991, 9.

[37] Relatório da Delegação do Brasil: À Conferência das Nações UNIDAS Sombre o Meio Ambiente, 1972, Itamaraty, Government of Brazil, 49.

environmental institutions at UNSSC in 1982, it was clear that the existing institutional setup had limitations. These limitations contributed to a mixed record in the implementation of the Stockholm action plan and to the frustration of efforts to ameliorate global environmental problems.

Two main political dynamics influenced the institutional design of the United Nations Environment Programme: the political and procedural requirement of gaining unified North-South backing and the necessity of maintaining the goodwill and concurrence of the other UN agencies. UNEP's efficacy relied heavily on a strong secretariat, a well-financed voluntary Environment Fund, and fruitful collaboration on the Governing Council. Equipoise and harmony between these elements was required for UNEP to function effectively. As we shall see, however, crosscutting interest cleavages did much to restrict UNEP's impact on the environmental problems that it was created to address.

From the first session of the UNEP Governing Council in June 1973, the delicate political consensus forged at UNCHE began to come apart. As one participant recalled, though the 1971 Founex meeting helped bring agreement, "the cracks were thinly papered."[38] At the first Governing Council session, the UNEP secretariat's draft programme – which aimed to give effect to the Stockholm action plan – was rejected outright by developing countries as insufficiently oriented toward their concerns.[39] As a result, many observers began to worry that the North-South divide on environmental questions was too wide to bridge. UNEP's Executive Director, Maurice Strong, began to speak of resigning. A fractious and intractable deadlock persisted until the final day of the two-week Governing Council session, when a text was cobbled together at the last possible moment by Britain's Dr. Martin Holdgate and Iran's Taghi Farver.[40]

Although UNEP had an approved action plan, tremendous damage had been done. Key developed country governments were left asking fundamental questions about the viability of a UN environment programme. Meeting in January 1974, leading industrialized states, collaborating in a format known as the "Brussels Group," reflected on the

[38] Holdgate 2003, 200. [39] Caldwell 1990, 72.

[40] Holdgate 2003, 201. Dr. Martin Holdgate Interview with the author, August 6, 2012.

first Governing Council session, which "was a near disaster." Despite "real doubt as to whether a successful environmental program within the UN context could get off the ground," recorded the meeting's Swedish participant, those present resolved to "try to make the second [Governing Council] session in Nairobi enough of a success to justify to their governments' continued participation in UNEP."[41] Thus, the very existence and continued operations of the new UN body were jeopardized from its earliest days. The governments that were expected to provide the main political and financial support for UNEP began to wonder if the programme could meet their basic goals.

While the second and third sessions of the Governing Council were less tumultuous and UNEP survived, political competition hindered greatly the new programme's efficacy. In its first decade, the UNEP Governing Council concerned itself almost exclusively with administrative matters and with the approval of the Environment Fund budget. As a consequence, the Governing Council failed to establish itself – as intended at Stockholm – as an authoritative intergovernmental body for environmental policy guidance, review, and coordination.[42] Instead, it served as an organization primarily concerned with brokering compromise between developed and developing countries on the uses of the Environment Fund. Industrialized states, which provided the bulk of the funding, were left "grasping towards a package in the Action Programme in which [they] would have to accept some projects with a distinct development flavour in order to secure some of the projects dear to [their] heart."[43]

Political competition and obsession with financial matters hampered the Governing Council's ability to serve as an effective means of coordinating UN environmental activities. Because of national bureaucratic competition, moreover, governments were unable to voice consistent positions in the different intergovernmental fora concerned with the many dimensions of world environmental management.[44]

41 "Memorandum: A Proposal for Communication – In the Context of the Second Meeting of the UNEP Governing Council at Nairobi in March," January 22, 1974, Riksarkivet, Statsdepartementens gemensamma organ, SE/RA/1200/07/F/F 7/4.

42 Ågesta Group AB 1982, 14.

43 Burton to Meg Rothwell, December 14, 1973, UK National Archives, FCO 55/1007 SME6/522/3 Part H.

44 Thacher 1991, 32.

Environment ministries – which were the most likely to serve as supportive constituencies to advance UNEP's aims within governments – were, as one writer put it, "the neglected children in governmental hierarchies." UNEP plans and priorities thus received relatively short shrift in the cabinet meetings of national governments.[45] "The inevitable conclusion," as a Swedish diplomat recalled, "was that as long as governments were not willing or capable of coordinating their own environmental policies in the UN system, the specialized agencies would continue to be largely independent actors supported by the different interest groups within countries."[46] The specialized agencies had state sponsors on the Governing Council that advanced their proposals and aimed to prevent the articulation of a more centralized vision.[47]

The UNEP secretariat was mandated by the General Assembly to "serve as a focal point for environmental action and co-ordination within the United Nations system."[48] As an organ without an operational mandate itself, UNEP was designed to function as the environment programme for the *entire* UN system, rather than as a sectoral body.[49] "At Stockholm and for a few years after," Robin Clarke and Lloyd Timberlake wrote in 1982, "'the programme' was understood as the sum of what all UN agencies, all governments and non-governmental bodies were doing about the environment. UNEP was seen as catalyzing and coordinating the 'the programme,' not implementing it."[50] It was a "switchboard" organization, facilitating the actions of other agencies. UNEP was not meant to be to the environment, for example, what the World Health Organization was to health. This mandate was widely misunderstood by suspicious UN specialized agencies aiming to protect their own sprawling bureaucratic fiefdoms.[51] UNEP's tiny size, its institutional weakness relative to other actors in the UN system, and the fact that it had few incentives to offer partner agencies constrained its ability to fulfill its

[45] Shabecoff 1996, 50. [46] Engfeldt 2009, 103.

[47] A British memorandum notes that: "The Specialized Agencies are exerting pressure on UNEP through developing countries to support their proposals at the Governing Council. UNEP's coordinating role will be in jeopardy if it is obliged to yield to these pressures." See: "Consultations with Developing Countries," 1974, UK National Archives, FCO 55/1007 SME6/522/3 Part H.

[48] UNGAR 2997 1972.

[49] United Nations Environment Programme March 24, 1976, 13–14.

[50] Clarke and Timberlake 1982, 49–50. [51] Shabecoff 1996, 48.

coordination role.[52] The unwillingness of specialized agencies to play their part in realizing UNEP's charge contributed to dismal political relations between the new programme and major UN bodies through much of the 1970s. Interagency relations became so frayed that tensions did not go unnoticed by states.[53] UNEP, moreover, enjoyed limited support in its early years from New York. Strong did not have a good relationship with UN Secretary-General Kurt Waldheim, who sought to have him replaced just as the new UN body was getting off the ground. Almost four decades later, the UN's bureaucratic politics had apparently left an impression on UNEP's first executive director. Indeed, in his recollections of UNEP's early years, Strong put it bluntly: "Waldheim was not a credit to the organization."[54]

UNEP's coordination difficulties were manifested visibly in the fate of the United Nations Environmental Co-ordination Board (ECB), a body that was formed to ensure "co-operation and co-ordination among all bodies concerned in the implementation of environmental programmes." To place the environmental secretariat at the center of this process, UNEP's executive director was mandated to chair the interagency ECB.[55] Less than four years after launching it, however, governments were forced to acknowledge that the ECB had been "one of the more difficult aspects of UNEP's role."[56] This was because the heads of UN agencies largely ignored it.[57] In 1977, the ECB was abolished and its responsibilities were folded into the UN's Administrative Committee on Coordination. The death of the ECB served as a "conspicuous sign" that interagency coordination was failing.[58]

Many of UNEP's early difficulties were compounded by its location in Nairobi.[59] Owing to the UNEP executive director's exacting travel schedule, one British official quipped in November 1973 that attempting "to come to grips with UNEP in Nairobi is like trying to measure a cloud of smoke." The fact that UNEP's senior staff spent so

[52] McCormick 1991, 112.
[53] United Nations Environment Programme February 16, 1976.
[54] Maurice Strong Interview with the author, August 20, 2010.
[55] UNGAR 2997 (XXVII) 1972.
[56] United Nations Environment Programme February 16, 1976.
[57] Munro 1991, 22. [58] Brenton 1994, 259.
[59] For an extended discussion of the implications of UNEP's location for its activities, see: Ivanova 2021, 117–139.

little time in Nairobi, the British memorandum continued, led to the impression that "UNEP is *in* Nairobi, but not *of* Nairobi."[60] The prolonged absence of staff from headquarters contributed to management problems. A blunt internal memorandum to Maurice Strong from R. A. Frosch, the Assistant Executive Director of UNEP, summarizes these problems:

> The demands of travel on you, Maurice, have been so great that the staff believes, and I am afraid with some reason for belief, that you are detached from the Organization, don't really know its views or what is being developed inside it, and are going around the world running your own piece of the programme, having nothing to do with what is being developed in the House; then you come back and try to re-impose your views.[61]

Indeed, in an interview, Strong confessed that he had not believed that the choice of Nairobi was a sound one for an organization focused on coordination and catalysis.[62] In addition to a disjuncture between staff and "the twenty-sixth floor" executives, UNEP also faced difficulties in recruiting and retaining qualified staff.[63] Consequently, its reports and draft decisions were considered by governments to be of variable quality.[64] In addition, as one interviewee remembered, "intellectual life there, then and now, is poor. It was not stimulating."[65]

UNEP was far from the agencies it was meant to coordinate. Many UN agencies would not assume the travel costs and time involved in sending officials to Kenya. The distance was, acknowledged governments, a "principal constraint" to UNEP's coordination mandate.[66] It was also a major obstacle to attracting senior officials and technical staff to important meetings, such as the annual Governing Council sessions. At the 1981 meeting of the UNEP council, for instance, less than 20 percent of delegates from member and observer countries were from departments with substantive responsibilities for environmental

[60] M. P. V. Hannam to W. N. Wenbam-Smith, "Possible Geneva Group in Nairobi," November 21, 1973, UK National Archives, FCO 55/1008 SME 6/522/3/Part J.

[61] R. A. Frosch, "Notes on UNEP," July 25, 1975, Environmental Science and Public Policy Archives, Harvard College Library, Peter S. Thacher Papers I, Box 117, Folder 1110.

[62] Maurice Strong Interview with the author, August 20, 2010.

[63] McCormick 1991, 111. [64] Engfeldt 2009, 105.

[65] Lars Björkbom Interview with the author, July 24, 2012.

[66] United Nations Environment Programme February 16, 1976, 14.

problems and priorities.[67] The developing countries were some of the least likely to send senior delegates from capitals. At a May 1975 informal intergovernmental meeting there was a near revolt among developing countries – led by the Central African Republic, speaking on behalf of the African group – when future UNEP consultations were scheduled for Nairobi. Because of the difficulties experienced in getting qualified technical and political staff to Kenya, these countries recorded their reservations.[68] As we shall see in Chapter 6, the inability of developing countries to field high-quality delegations in Nairobi also badly damaged UNEP's status during preparations for the 1992 UN Conference on Environment and Development.

Developing states, which were strongly represented on the UNEP Governing Council, were in a position to influence heavily the use of the Environment Fund. The problem with this arrangement, as far as the industrialized countries were concerned, was that they were the ones expected to make sizable voluntary contributions to the fund. UNEP's financial viability, therefore, rested upon a setup where one set of countries was expected to contribute to a fund that was controlled mainly by another set of countries. While donors were able to exercise influence on the use of funds because it was hoped that they would contribute in the future, they were pressured to consent to expenditures on projects that were not of primary interest to them.[69]

After the initial wave of financial commitments – made in the context of the Stockholm conference – was met, industrialized countries' contributions to the Environment Fund began to flag.[70] An October 1973 British memorandum is illustrative of the cynicism and disillusionment that quickly emerged among some industrialized countries over contributions to the fund. There was, the report noted, "no mileage to gain by becoming involved or committed where our interests were academic," such as in anti-desertification, sanitation, or

[67] Ågesta Group 1982, 15.

[68] "Unofficial Record of Informal Consultations of Representatives of Regional Groups and Members of the Secretariat of UNEP, Kenyatta Conference Centre, Nairobi," May 1, 1975, Riksarkivet, Statsdepartementens gemensamma organ, SE/RA/1200/07/F/F 7/10.

[69] Burton to Meg Rothwell, December 14, 1973, UK National Archives, FCO 55/1007 SME6/522/3 Part H.

[70] For a more extended discussion of UNEP's funding challenges, see: Ivanova 2012.

human settlements.[71] After deploring the fact that "signs were clearly emerging ... of a determination by certain states to exploit UNEP as a potential source of development aid," the Foreign and Commonwealth Office recommended to the UK cabinet that its contributions to the fund should be "just sufficient" to "exercise influence and maintain credibility."[72] In 1976, the US$40 million pledge of the United States to the fund was reduced to US$25 million.[73] The Governing Council's approved annual budget was US$37.5 million per year for 1978–1981, but UNEP could only raise about US$31 million per annum.[74] UNEP's budget began to diminish in real terms by the late 1970s and early 1980s. Set against inflation, UNEP's 1982 budget had fallen to US$15.6 million per year in 1973 dollars – a total significantly lower than the target set by states in Stockholm.[75] Every year, remarked Dr. Mostafa Tolba, who took over from Maurice Strong as UNEP's Executive Director in 1975, UNEP was forced to "exercise the most painful prudence in committing funds for new and important activities which the Council has earlier approved but for which sufficient resources have not been made available. Year after year, our hands have been tied."[76]

In addition to a stagnating budget, UNEP suffered from late or erratic contributions from governments, an unwillingness from some states to commit to funding levels more than a year in advance, and a large percentage of funds coming in nonconvertible currencies. All of these factors meant that even the meager resources made available to UNEP could not be utilized in a straightforward manner. Since the Environment Fund was designed to serve as a means of carrying out the Stockholm action plan by catalyzing action in the UN system, UNEP's limited financial capacity constrained its ability to induce the

[71] "Minutes of a Meeting Held in Room N 19/21, Wednesday 31 October 1973 at 10:30 a.m.," UK National Archives, FCO 55/1007 SME6/522/3 Part H.

[72] "United Nations Environment Programme – U.K. Policy," Draft Cabinet Paper, November 1973, UK National Archives, FCO 55/1007 SME6/522/3 Part H.

[73] Foreign and Commonwealth Office, "32nd Session of the United Nations General Assembly: United Nations Environment Programme," ICC (77), UK National Archives, October 1977, FCO 58/1195/UNS 261/5 Part B.

[74] McCormick 1991, 110.

[75] Goldsmith 1982.

[76] Tolba 1982a, 141. For an extended discussion of UNEP's finances, see: Ivanova 2021, 39–44.

cooperation of other UN agencies.[77] A significant proportion of UNEP's annual expenditures, furthermore, was sapped because it was obliged to support the ongoing projects of specialized agencies that refused to assume the full financial burden of initiatives launched in partnership with UNEP years earlier.[78] These agencies – some of the largest in the UN – enjoyed political backing from developing countries.[79] UNEP's uncertain budgetary footing may have also hampered its ability to produce environmental assessments that were critical of member states.[80]

UNEP's difficulties in effectively coordinating and catalyzing larger, well-established UN specialized agencies meant that states' belief that a small body could serve as an effective point of focus for the activities of the UN system had to be reassessed. As UNEP's first decade ended, it was obvious to all concerned that UN environmental institutions had important limitations. Swedish diplomat Lars-Göran Engfeldt recalled: "By 1982, it was completely clear that the specific features built into UNEP – the Environment Fund and the special position of the [Executive Director] – had not been enough to compensate for the basic weakness of the overall post-Stockholm institutional arrangements."[81] In addition to flaws in the UN framework for environmental cooperation, the task of brokering North/South compromise was made more difficult by the need to balance short and long run economic priorities and to address a subject of great technical and scientific complexity.[82]

All of these factors contributed to a mixed record of UN environmental action from 1972 to 1982. An extensive review of the implementation status of the Stockholm action plan was conducted in 1982 by the Ågesta Group, a Swedish parliamentary initiative. The

77 The UNEP Governing Council recognized by its second session that the Environment Fund would be insufficient to fund "wholly or partly" all new UN environmental initiatives. See: Ågesta Group 1982, 18. See also: McCormick 1991, 112.

78 Munro 1991, 9.

79 "United Nations Environment Programme – U.K. Policy," Draft Cabinet Paper, November 1973, UK National Archives, FCO 55/1007 SME6/522/3 Part H.

80 Edward Goldsmith, a UNEP insider, noted in 1982: "Clearly UNEP is muzzled. It can only say things that do not offend member states and can only do things that do not run counter to their political and economic interests." See: Goldsmith 1982.

81 Engfeldt 2009, 103.

82 Thacher 1991, 22.

group concluded that, of the 109 recommendations and more than 150 distinct proposals contained in the UNCHE action plan, only 30 percent had been implemented. Forty percent were classified as partially implemented, 15 percent had received little or no attention, and vague recommendations made it difficult to evaluate the final 15 percent.[83] Important highlights of the implemented items included recommendation 74(e), which proposed an International Registry of Data on Chemicals in the Environment; 99(i)(a), which led to the World Heritage Convention; recommendation 99(i)(b), which stimulated the completion of the Convention on Wetlands of International Importance; recommendation 99(3), which formed the basis for the Convention on International Trade in Endangered Species of Wild Fauna and Flora; and recommendation 101, which led to the creation of a International Referral Service.[84]

The recommendations that were classified as partially implemented either saw part of the proposed measure enacted, experienced an unsuccessful result or were initiated but not pursued. Other recommendations were almost completely ignored.[85] In fairness, the informational basis for environmental action matured dramatically in the decade after UNCHE. Appreciations of the state of the world environment were revolutionized in the 1970s by such factors as satellite technology and cheap electronic data storage and retrieval systems.[86] These developments led to an increased awareness of the inadequacy of the state of knowledge about the world environment, and motivated members of the scientific community to point out the need for international cooperation to facilitate better data collection and analysis.[87]

By 1982, it was clear that the post-Stockholm institutional arrangements had faced important challenges in implementing the UNCHE action plan. Instead of enabling a coordinated UN approach to environmental exigencies, UNEP succumbed to crosscutting North/South divisions and an inability to bring coherence to the fragmented activities of UN specialized agencies. Designed in 1972 to remedy inefficiency, duplication, and overlap in UN environmental institutions, UNEP had made limited progress.

[83] Ågesta Group 1982, 7. [84] Ågesta Group 1982, 8.
[85] Ågesta Group 1982, 9–10. [86] UNEP 1982, 9–10.
[87] Holdgate 1979, 178–181.

The 1982 Nairobi Conference as a Temporal Focal Point

When states gathered in the Kenyan capital in May 1982, they focused on a mixed post-UNCHE implementation record and an institution that faced challenges in meeting the ends that they had established for it just ten years earlier. Before discussing the institutional bargaining that targeted these problems, I will show that the 1982 meeting served as a Temporal Focal Point in UN environmental politics.

The UNEP Session of a Special Character was convened in accordance with General Assembly resolutions 35/74 of December 5, 1980, and 36/189 of December 17, 1981.[88] The May 10–18, 1982 conference, held in Nairobi, aimed to commemorate the 1972 UN Conference on the Human Environment. The session was attended by 105 states – a total similar to the 113 that attended the Stockholm conference – and included several heads of state and government. Like the Stockholm conference, the major outputs were a declaration and a plan of action. In addition, four resolutions were adopted, including one on the need to launch a special commission on long-term environment strategies.

Though its results were ultimately disappointing to reformers, the session served as a Temporal Focal Point in the history of UN environmental affairs. For a discrete phase of the temporal continuum to be considered focal, it must be unique and highly conspicuous. As with Stockholm, the fact that the Nairobi conference represented a discrete point in time is uncontroversial. The conference was a distinct, nine-day period in time. States knew of the session's existence at least since February 1980, and those that wished to advance initiatives in the context of the approaching meeting did so.

The main rationale for the timing of the special session was the fact that the landmark Stockholm conference had occurred a decade earlier. UN General Assembly resolutions 35/74 and 36/189, which proposed and made arrangements for the UNSSC, cited the tenth anniversary of the Stockholm conference as a chief rationale for scheduling the meeting.[89] By invoking the legacy of the Swedish conference, resolution 36/189 called on states to meet the "need to revive the sense of urgency and commitment by Governments for national and

[88] UNGAR 35/74 1980. [89] UNGAR 35/74 1980; UNGAR 36/189 1981.

international co-operative action to protect and enhance the environment."[90]

The proposal for the Nairobi conference came primarily from the UNEP secretariat, rather than a particular government. In fact, some prominent states quietly questioned the wisdom of the decision to launch the UNSSC, wondering if UNEP's executive director had thought through all of the legal, organizational, and financial implications of the initiative.[91] While the General Assembly resolutions pointed to the worsening state of the world environment as a primary motivation, they were not specific about the particular problems that the UNEP session targeted. No institutional or environmental triggers provide a compelling explanation for the timing of the Nairobi conference. The UNCHE anniversary was needed to provide the rationale for a major gathering. The gradual deterioration of the global environment and the organizational limitations of UNEP were by themselves insufficient justifications.

The special session was both unique and highly conspicuous. It marked the anniversary of a transformational event in the international system. UNEP's executive director aimed deliberately to increase the profile of the Nairobi conference in key countries. In a manner reminiscent of Maurice Strong's efforts to focus attention on the Stockholm conference, Tolba traveled to key countries, delivering speeches and giving interviews.[92] In his pre-UNSSC speeches, the Egyptian microbiologist highlighted the pressing nature of global environmental problems and argued that the Nairobi conference presented an opportunity to reevaluate institutional arrangements.

In November 1981, Tolba spoke to the Environment Committee of the European Parliament about the "urgent need for recommitment" to the environmental cause in time for the approaching Nairobi conference. The anniversary of Stockholm should, he noted, focus governments' minds on the global environmental agenda. He encouraged those present to reevaluate institutional arrangements, which, he noted bluntly, were a bit of "a thankless brief."[93] In January 1982, Tolba

[90] UNGAR 36/189 1981.

[91] US officials, for instance, questioned whether Mostafa Tolba had thought through all of the legal, financial, and organizational implications of the session. See: J. E. Roberts (UKDEL-NAIROBI) Memorandum, February 6, 1980, UK National Archives, FCO 58/2186, UNS 261/7.

[92] Lamb 1982, 1. [93] Tolba, 1982d.

spoke at the Japan Environmental Agency symposium, referring to the UNSSC as "a once-in-a-decade chance" to address international environmental problems. To add a sense of urgency, he discussed the ominous array of international environmental problems confronting the world community and concluded by stressing that the Nairobi conference was a "unique opportunity" for the international community to tackle these issues. It was time, Tolba stressed, for Japan to play a lead part in "making the world at the turn of the century a safer and more prosperous one than we find it today."[94] This message matched the tone and emphasis that Tolba communicated to states at the ninth session of the UNEP Governing Council.[95]

In 1980, Tolba commissioned an extensive report, titled *The World Environment 1972–1982*, edited by Martin W. Holdgate, Mohamed Kassas, and Gilbert F. White.[96] The report's purpose was to establish UNEP's capacity for environmental assessment, a dimension of its mandate that had not been especially prominent in its first decade.[97] The contribution established a substantive basis for the Nairobi conference. A shorter – perhaps more government friendly – version of the Holdgate group's findings was published by UNEP as *The State of the Environment, 1972–1982.*[98] In addition, the secretariat produced two other reports to motivate action at the Kenyan meeting: *The Environment in 1982: Retrospect and Prospect* and *Major Environmental Trends to be Addressed by UNEP, 1982–1992*. Thus, to add to the conspicuousness of the meeting and the preparedness of states, the UNEP secretariat produced an impressive array of technical and popular reports, meant to appeal to public opinion and to government officials alike. These reports came in all shapes, sizes, and styles. The UN also produced a set of commemorative stamps to drum up excitement for UNSSC.[99]

Though the UNEP secretariat's travels and public diplomacy played an important part in raising the profile of the Nairobi conference, the session and the tenth anniversary of UNCHE caught the eye of many other actors independent of Tolba's efforts. The focus on the 1982 conference as a potential vehicle for rejuvenating the UN environmental enterprise was emphasized in a January 1981 United States Council on

[94] Tolba 1987, 1–7. [95] UNEP 1981. [96] Holdgate et al. 1982.
[97] Holdgate et al. 1982. [98] United Nations Environment Programme 1982.
[99] Tower 1982, 2.

Environmental Quality and State Department report to the president. *Global Future: Time to Act* argued that "the 1982 UNEP Conference should be expanded to be more than a ceremonial occasion," focusing attention on key environmental issues and commencing "a fresh evaluation of UNEP's role and how it could be supported and improved."[100] In March and April 1982, the US House of Representatives Subcommittee on Human Rights and International Organizations held hearings for the "Review of the global environment 10 years after Stockholm." The congressional hearings used the Stockholm anniversary as a reason to assess US foreign environmental policy heading into the Nairobi TFP.[101] In April 1982, just in time for the conference, the European Economic Community met to reconsider the role of ecological imperatives in their development assistance projects.[102] The Swedish government, furthermore, regarded the Nairobi conference as "a suitable time to put UNEP's future and direction on the agenda."[103] High-level environmental hearings examining transnational environmental problems were held in London in June 1982.[104]

The Nairobi TFP also stimulated the production of literature from governmental and nongovernmental sources that aimed to influence opinion surrounding the Nairobi conference. The Ågesta Group report, *Twenty Years After Stockholm, 1972–1992*, was circulated widely to conference participants. The Ågesta report was heavily based on the observations of "Stockholm veterans."[105] The International Institute for Environment and Development's Robin Clarke and Lloyd Timberlake composed *Stockholm Plus Ten: Promises, Promises? The Decade Since the 1972 UN Environment Conference.* Clarke and Timberlake's contribution summarized official documentation in a media-centric manner and provided an extensive assessment of the state of the global environment in 1982.[106] Erik P. Eckholm's *Down to Earth: Environment and Human Needs* was a Nairobi

[100] United States Council on Environmental Quality and Department of State 1981, 184–185.
[101] United States Congress – House Committee on Foreign Affairs, Subcommittee on Human Rights and International Organizations 1982.
[102] Cooney 1982, 8.
[103] Lars Björkbom Interview with the author, July 24, 2012.
[104] Lean and Mayer 1982, 2. [105] Ågesta Group 1982.
[106] Clarke and Timberlake 1982.

equivalent of Barbara Ward and René Dubos' *Only One Earth* and had an impact on popular opinion.[107]

While the 1982 Nairobi conference has largely been forgotten in the literature on global environmental governance, it was therefore a highly exceptional event at the time. Recalling the Stockholm conference, this exceptionality was deliberately fostered by the secretariat through a series of speeches, interviews, and publications. The secretariat's job was eased by the fact that the tenth anniversary of the Stockholm conference provided a compelling focal point for states to evaluate global environment change and to consider the post-Nairobi UNEP action program.

The 1982 Nairobi conference thus served as a Temporal Focal Point in the history of UN environmental institutions. It offered an exceptional and discrete period in time for governments to focus on the world environment and on the status of post-Stockholm institutional arrangements. The uniqueness and conspicuousness of this point on the temporal continuum flowed mainly from the anniversary of UNCHE. Reform-minded governmental and nongovernmental bodies were drawn to UNSSC as a potential vehicle for action. By giving rise to a temporal convergence of expectations, the focal phase helped to stimulate numerous programmatic and institutional initiatives in Nairobi.

Nairobi and the Absence of Joint Gains

When states gathered in Nairobi, their attention was focused on the record of world environmental action between 1972 and 1982. The Stockholm action plan had only been partially implemented and important environmental issues had been left un(der)addressed. UNEP's mixed performance was the result of numerous factors, including limited funding, an inability to coordinate the activities of larger UN agencies, and the programme's administrative location in Nairobi. To meet UNEP's limitations, Kenya and Tunisia, backed by the African group, argued that UNEP's mandate should be widened so that it could undertake operational activities. Their proposal failed to generate support and was ultimately defeated. Although UNSSC provided a focal timeline that helped elicit institutional initiatives,

[107] Eckholm 1982.

underlying conditions had not changed greatly in the decade since the Stockholm conference and provided no basis for the realization of joint gains. This section will provide background on the African group proposal and will analyze the reasons for its failure.

At the Nairobi conference, Kenya and Tunisia pushed energetically for UNEP to gain an operational mandate. In his conference address, the Kenyan President, Daniel arap Moi, noted that, when "UNEP was first conceived, it was projected as something like a catalyst, knitting together elements of initiative into a world-wide compound of vital consequence. But some of us have always felt that this approach was over-cautious, and would much prefer to see a more fearless performance." Moi urged governments and UNEP to move beyond "talking in comfortable arenas," to action, "taking coats off and getting honourably dirty."[108] Kenya's representative to the conference plenary urged the environment programme to act on the control of water-borne diseases, the implementation of a renewable energy plan of action, and to initiate steps against environmentally deleterious pest control.[109] Agreeing, Helen Benitez of the Philippines asserted: "It is our position that this special session should set the direction for a rethinking of the institutional role of UNEP. The organization must be strengthened, maybe even restructured, to enable it to become more responsive to the decade ahead ... The prospect for success in global environment issues will be bleak unless we are able to bring about certain major changes."[110] Though welcoming the need for more research, Egypt's representative observed that academic studies "seem to overshadow the real services to protect the environment."[111]

The provision of resources to developing countries was critical to meeting the dual ends of environmental preservation and development. Representatives from Côte d'Ivoire stressed the self-interest of developed countries in assisting developing states in addressing their environmental needs, since "in the long and not so long term, all mankind, in one way or another, will have to suffer the consequences of pollution or environmental degradation, the destruction of nature and the depletion of natural resources."[112] Kenya also called on

[108] Moi 1982, 8–10. For the official account of this debate at UNSSC, see: United Nations Environment Programme A/37/25, 14.

[109] "Kenya, Nairobi, 1982," Tolba 1982e, 244–245.

[110] Benitez 1982, 25.

[111] "Egypt, Nairobi, 1982," Tolba 1982e, 171.

[112] "Ivory Coast, Nairobi, 1982," Tolba 1982e, 230–231.

developed countries to "mobilize the necessary resources to assist developing countries to achieve an improved level of self-reliance in integrated environmental management."[113] A transfer of wealth from North to South and an endorsement of the precepts of the New International Economic Order were, it was maintained, essential to safeguarding the global environment.

One possible means of implementation would be to give effect to a Swedish initiative to open a special "development window" in the utilization of the Environment Fund.[114] The development window, discussed extensively at UNEP's ninth Governing Council session in 1981, would allow funds to be allocated to the gravest environmental problems of developing countries, as identified by the Governing Council. Swedish officials argued that the window could be used to fund environmental impact assessments before launching large development projects. It could also facilitate fuller participation by developing countries in global environmental endeavors.[115]

The Kenyan/Tunisian proposal would alter UNEP's mandate by orienting it increasingly toward projects that were essentially national or regional in scope. Initiatives could be launched on a country-specific basis, they proposed, shifting further the character of UNEP's agenda toward the concerns of developing countries. As emphasized in the Malawian speech to the UNSSC plenary, the environmental concerns of developing countries "are essentially of a national nature."[116] The lack of resources and government capacity were major impediments to environmental management in the nonindustrialized world.[117] Inadequate information, a shortage of trained specialists, lack of funds, and obsolete equipment were important obstacles to environmental policies that, in the absence of greater international financial and technical assistance, could not be overcome in the near term.[118]

A mandate shift was backed by Tolba, who, we have noted, traveled the world in 1981 and 1982 providing a rationale for a UNEP role in project implementation. After underscoring the difficulties and

[113] "Kenya, Nairobi, 1982," Tolba 1982e, 246.
[114] United Nations Environment Programme A/37/25, 17. Engfeldt 2009, 105.
[115] United Nations Environment Programme 1981, General Debate, paragraphs 438–446.
[116] "Malawi, Nairobi, 1982," Tolba 1982e, 263.
[117] "Kenya, Nairobi, 1982," Tolba 1982e, 246.
[118] Leonard and Morell 1981, 300–301.

limitations associated with UNEP's existing mandate, UNEP's executive director invited the governments of the European Parliament to assess the efficacy of UNEP's catalytic role at UNSSC.[119] In Tokyo, Tolba extolled the virtues and foresight of industrial states' investment in environmental protection in the global South: "it pays for rich nations to invest in the poor. These are not the politics of the begging bowl, but of international justice and benign self interest."[120]

The proposed changes to UNEP's mandate implied a significant alteration to the North/South environmental bargain of the Stockholm conference.[121] Change would address two main limitations of the UN programme. First, by enabling UNEP to implement projects itself, it would be less reliant on coordinating uncooperative specialized agencies. The programme's limited influence in the UN system would be less of an impediment if it had its own operational mandate. Second, it would increase the relevance and visibility of UNEP in the developing world.[122] By making UNEP more pertinent to the concerns of the global South, states could realize the type of global ethic envisioned at Stockholm.[123]

Despite these efforts, the initiative was defeated. It ran into stiff opposition from Latin American countries, which were concerned that an operational role for UNEP could invite unjustified intrusions on their sovereignty.[124] Most industrialized countries argued that, given the limited resources available in the Environment Fund, UNEP should avoid a dispersal of effort, focus mainly on international problems, and refrain from direct aid schemes or executing agency responsibilities.[125]

[119] Tolba 1982a, 186. [120] Tolba 1987, 1.

[121] This followed an action orientation that was increasingly called sustainable development. The term "sustainable development" was first proposed by the 1980 World Conservation Strategy. Its conceptual antecedents, however, were the 1974 Cocoyok declaration and the report of the Brandt Commission. See: "The Cocoyoc Declaration: Adopted by the Participants in the UNEP/UNCTAD Symposium on 'Patterns of Resource Use, Environment and Development Strategies.'" Cocoyoc, Mexico, October 8–12, 1974. IUCN: 1980; The Brandt Commission 1980.

[122] Engfeldt 2009, 105.

[123] Commenting on this shift, Barbara Ward wrote: "Now, in the 1980s, some Western leaders are starting to abandon the concept of our joint voyage on Spaceship Earth, and to dismiss any concern for the environment or for the South as 'do-goodism.' They regard the universe as expendable – as long as their tiny bit of it is not expended." See: Ward 1982, 26.

[124] Engfeldt 2009, 150.

[125] United Nations Environment Programme 1982, 16.

Efforts to pare down UNEP's activities to "those environmental problems of the highest priority and in those areas where its influence can be decisive" had been a long-standing aim of leading developed countries.[126] The result of these conference debates was the conclusion at the Nairobi conference that existing institutional arrangements were "generally speaking, adequate and appropriate."[127]

Negotiations in Nairobi, therefore, did little to address UNEP's organizational limitations. UNEP's traditional role as a catalyst and coordinator was maintained. No significant new political commitments were made, and the Nairobi declaration and action plan was stuffed with platitudes and standard invocations. Commenting on the conference, Philip Shabecoff of the *New York Times* lamented that the "second United Nations Conference on the Global Environment, like a second honeymoon, was less exciting than the original."[128] It was, according to another observer, an opportunity that was "unfortunately squandered."[129]

How do we account for the failure of the Nairobi conference to bring about institutional change, even in the presence of a temporal convergent of expectations triggered by a TFP? As elaborated above, the challenges facing UN environmental institutions were well-known and widely accepted. In case there were any misconceptions, Mostafa Tolba told the conference plenary: "The inevitable consequence is that the fundamental objective of Stockholm to protect and enhance our environment for future generations has not been fulfilled."[130] Although preparations for UNSSC had been an "alarm bell," focusing attention on UNEP's problems, states did little more than acknowledge that the UNCHE action plan "has only been partially implemented, and the results cannot be considered satisfactory."[131] Despite this admission, states failed to reform an organization that had not met their purposes as set out in the Stockholm conference action plan.

Efforts to bring about institutional change in 1982 failed because cooperative conditions had not shifted greatly since 1972. There had been no major international economic or technological developments

[126] Speech by the United Kingdom Delegation, "Coordination Questions," April 23, 1980, UK National Archives, FCO 58/2186 UNS 261/7.

[127] United Nations Environment Programme 1982, 39.

[128] Shabecoff 1982c, 183. [129] Munro 1991, 9. [130] Tolba 1982b, 6.

[131] United Nations Environment Programme 1982, 49. The reference to UNSSC as an alarm bell is made in: Engfeldt 2009, 107.

that altered the cooperative setting in a manner that rendered existing institutions suboptimal. The distribution of power in the international system had not shifted in a way that could influence UN environmental politics. Most of the environmental concerns facing states at UNSSC had, moreover, been on the agenda at the Stockholm conference. As one observer noted, in 1982 there was no "substantive evidence that the system was in danger of imminent collapse."[132] The institutional arrangements adopted at Stockholm reflected these conditions and, thus, there was little incentive to alter cooperative structures a decade later.

In fact, better information on environmental conditions led certain problems that seemed dire in the early 1970s to be largely dismissed in the early 1980s. While still a concern, for example, estimates of the rate of tropical deforestation gathered from satellite imaging in the 1980s were more modest than the rate inferred from the ground observations of the 1970s. In a similar way, advances in scientific methods caused marine scientists to revise many of their earlier estimates about concentrations of trace substances in the world's waters.[133] There was far greater awareness of the limitations of climate modeling in the early 1980s, contributing to more modest conclusions about the rate of global warming.[134] Popular computer-generated scenarios, such as those published by the Club of Rome in *Limits to Growth* in 1972, were widely rejected as alarmist.[135]

Although desertification continued to be an important problem, there were encouraging signs that progress was being made on many local and regional issues. The first post-UNCHE decade had seen some UN cooperation to preserve soils, including the production of the first ever world "soils map."[136] Outside of Africa, the rate of population growth in developing countries had slowed between 1975 and 1980,

132 Struthers 1983, 276. 133 UNEP 1982, 10.

134 Estimates of the rate of climate change were more cautious in 1982 than in 1972. The contrast between 1972's *Only One Earth* and 1982's *The World Environment: 1972–1982* is striking. Barbara Ward and René Dubos' noted near scientific consensus that human activities would have "profound and unforeseen effects on the earth's climate." In contrast, Martin Holdgate et al. talked extensively about "major uncertainties" in climate projections. See: Holdgate et al. 1982, 57, 62; Ward and Dubos 1972, 195. For further discussion, see: Clarke and Timberlake 1982, 25–26.

135 Meadows et al. 1972. 136 Clarke and Timberlake 1982, 16.

potentially reducing stress on earth systems.[137] UNEP's Regional Seas Programme had been a notable success, catalyzing cooperative efforts to tackle pollution in ten international bodies of water between 1975 and 1982.[138] Although implementation had been limited in some areas, international conventions on the trade in endangered flora and fauna, marine pollution, world heritage, and wetlands began to improve the world environmental situation.

Although UNEP did face important limitations, the organization was still relatively new in 1982 and, after a focused assessment, many actors believed that few definitive conclusions could be drawn concerning the suitability of its design. Many took a wait and see approach, therefore. UNEP's principal challenges derived from declining contributions to the Environment Fund and to an unwillingness among member states to force other UN agencies to play nice. These aspects of UNEP's record were, arguably, the most responsive to the priorities of member states and, far from reflecting a structural problem, states could readily scale-up their financial and political support as appropriate.

Stability in underlying conditions meant that there were few change incentives to be realized through altering institutional norms, rules, or procedures in 1982. Key parameters remained unchanged and the current institutional setup was superior to available alternatives at the time. Although the Nairobi TFP helped elicit the African initiative, there was no bargaining range that could form a basis for agreement among a large coalition of actors to recast UN environmental institutions. Institutional alternatives were not payoff-dominant and the status quo held.

Temporal Focal Points and the Bargaining Range

The Nairobi meeting served as a Temporal Focal Point in UN environmental affairs, generating a temporal convergence of expectations. The tenth anniversary of the 1972 Stockholm conference provided powerful impetus for institutional actors – governmental and nongovernmental – to reassess post-Stockholm institutional arrangements. The moment was unique and highly conspicuous. While the UNEP

[137] Clarke and Timberlake 1982, 15.
[138] Clarke and Timberlake 1982, 42–43. See also: Johnson 2012, 57–61.

secretariat of Mostafa Tolba did much to increase the profile of UNSSC, the Stockholm anniversary generated considerable attention independent of its endeavors. Despite this, the results of the Nairobi conference – especially when compared to UNCHE – were disappointing to reformers. No major political commitments were made and the most prominent initiative to reform UNEP was defeated. The failure to act is explained by a lack of exogenous change that could open up sufficient incentives to alter UNEP's institutional arrangements.

A temporal convergence of expectations is, therefore, an insufficient explanation for the arrival of institutional change. There are nevertheless reasons to believe that the Nairobi TFP initiated a process that increased the likelihood of change. First, UNSSC provided an important means of temporal coordination for institutional actors. The special session was seen by reform-minded actors as the most likely platform to generate support for far-reaching institutional change. It then produced numerous political initiatives that reflected an investment in diplomatic resources. Mostafa Tolba, through his many interviews and public speeches, acted deliberately to put the state of UNEP on government agendas. Kenya and Tunisia, furthermore, used UNSSC as a vehicle in an attempt to alter the North/South environmental bargain of the Stockholm conference. In the ninth UNEP Governing Council session, the Canadian delegation raised the possibility of forming an independent commission on long-term environment strategies at UNSSC. This proposal was picked up and championed by Japan at the Nairobi conference. All of these deliberate efforts were undertaken to use the tenth anniversary of UNCHE as a window to break the institutional inertia that had seemed to set in. Important institutional actors invested political and analytical resources in pursuit of their goals.

Second, the level of institutional and environmental analysis was heightened considerably in the lead-up to the Nairobi conference. Actors expended scarce analytical and human assets in the assessment of the global environment. Despite the ongoing frustrations of many actors with UNEP's operations, the wide literature produced in the lead-up to the conference did much to increase the quality of information and analysis available on the state of UNEP and of the global environment. Prominent speeches and governmental hearings contributed to an information-rich environment. The focal phase helped generate a proliferation of popular books and institutional diagnostics.

Third, though existing treatments suggest that the failure of institutional change at the conference was inevitable, this interpretation is misleading. The conference sparked a major reassessment of policy approaches in many important governments. The attitude of the United States was, for example, highly important to the success of reform efforts. While it is true that the abortive attempt by the newly elected US administration of President Ronald Reagan to cut all US funding to UNEP in 1982 did not portend well for prospects for institutional change, the approaching conference sparked a major reconsideration of American policy toward the organization. Following the conference, one US congressional staffer and conference participant reflected:

> it is unlikely that high-level Administrative officials would have focused on global environmental issues without the Session of a Special Character. The conference forced the Administration to address problems in this field. In addition, the Session of a Special Character provided a rather unique opportunity for new, senior officials in EPA and the Interior Department to gain first-hand exposure to the views of other developed and developing nations on international environmental problems.[139]

The conference served as a means of focusing the minds of leading US officials on the UN's environment programme.

The Nairobi conference triggered an intense interagency debate within the US government over the role of UNEP in US foreign policy. It is through the lens of these internal discussions that interests are clarified and bargaining positions are solidified. The US negotiating position at Nairobi was the subject of conflict between the United States Environmental Protection Agency (EPA) and the State Department. The Department of State, led by Under Secretary James L. Buckley, had taken a more positive view toward UNEP than Anne Gorsuch, the administrator of EPA.[140] Believing that UNEP served US interests, Buckley had intervened after the administration proposed to cut all funding to the organization and managed to gain support for a US$2 million contribution to the Environment Fund.[141] Congress later increased this total to US$7.85 million.[142] Yet despite their success in

[139] Grunberg 1982, 495.
[140] Gordon S. Brown Interview with the author, April 3, 2013.
[141] James L. Buckley Interview with the author, March 26, 2013.
[142] Shabecoff 1982a, A2.

helping to limit damage to UNEP, the State Department played a relatively minor role in preparing the US position at UNSSC.[143] At the conference, EPA dominated the US negotiating position and the State Department had minimal input into the content of Gorsuch's polarizing conference speech.[144] On the role of the State Department in US preparations, one of the lead US negotiators recalled in an interview with the author: "To be frank, there was a great deal of confusion [among State Department officials] at the conference on just what we were doing. We didn't know that Madam Gorsuch would be such a pain in the neck until she showed up."[145]

This is not to suggest that the US negotiating position would have been significantly different had the State Department led negotiations. The US position was largely determined by structural factors. It was the underlying conditions of the international system that rendered the status quo preferable to any available institutional alternative in 1982. What is essential to note, however, is that the Nairobi conference TFP focused the minds of decision-makers on US policy toward UN environmental cooperation. The fact that UNSSC gave rise to interagency jockeying and provided an opening for debate suggests that, if conditions had been different, this process might well have led actors to recognize the existence of change incentives. As one American participant reflected: "Things tend to have an inertial quality until a decision point at some point in time forces governments to reach a policy decision. Cabinet-level conferences, like the Nairobi meeting, sharpen the minds of people concerning particular agencies in foreign policy."[146]

Conclusion

After an introduction to the origins of UNEP and its politics, this chapter analyzed the fate of institutional change proposals at the 1982 United Nations Environment Programme Session of a Special Character. The Nairobi conference served as a Temporal Focal Point, producing a convergence of expectations among institutional actors. Despite institutional limitations, the unavailability of joint gains led to

[143] Gordon S. Brown Interview with the author, April 3, 2013.
[144] James L. Buckley Interview with the author, March 26, 2013.
[145] Gordon S. Brown Interview with the author, April 3, 2013.
[146] Gordon S. Brown Interview with the author, April 3, 2013.

the defeat of an African group reform proposal. This case study, therefore, supports the hypotheses contained in Chapter 2: The presence of a temporal convergence of expectations will not produce institutional change when international conditions do not provide a basis for joint gains.

Evaluation of the case study has two notable implications for our consideration of institutional change processes. Although the presence of a temporal convergence of expectations is an insufficient condition for change, the case study demonstrates the role of exceptional periods along the temporal continuum in focusing minds on the state of institutional arrangements. The Nairobi conference, for example, precipitated a vigorous internal debate in Washington on the state of the world environment and on the place of UNEP in US foreign environmental policy. Evidence suggests that it had a similar effect in other capitals around the world.[147] Hence, there is reason to believe that convergent expectations may have, in more favorable circumstances, caused actors to realize favorable agreement possibilities that had previously gone unnoticed.

In addition, the presence of institutional limitations is an inadequate explanation for the arrival of institutional change. Important limits in UNEP's design were evident to states as they convened in Nairobi in 1982. In the UNSSC declaration, governments acknowledged the disappointing record of the UN Environment Programme in implementing the Stockholm action plan. As will be suggested in Chapters 5 and 6, UNEP's many limitations were likely self-undermining in the longer run.[148] While UN environmental institutions remained self-enforcing through 1982, UNEP's limitations slowly eroded the range of parameters in which the institutional status quo was tenable.

[147] Engfeldt 2009, 105–107. [148] Grief and Laitin 2004.

5 *The Brundtland Commission and the Seeds of Change*

By its second decade, the United Nations Environment Programme (UNEP) had compiled a mixed record in arresting global environmental deterioration. UNEP had played an indispensable part in helping to precipitate the ozone accords, including the high-profile Montréal Protocol on Substances that Deplete the Ozone Layer. However, UNEP's principal limitations, acknowledged by states at the 1982 UNEP Session of a Special Character, were apparent. The Third World debt crisis, which became dire in the mid-1980s, exacerbated the environmental effects of important shifts in underlying conditions that rendered existing modes of cooperation suboptimal.

It was in this context that the World Commission on Environment and Development (WCED), chaired by Gro Harlem Brundtland, delivered its report to the United Nations General Assembly in autumn 1987. *Our Common Future* popularized the concept of sustainable development and advanced a series of far-reaching recommendations for institutional change.[1] Though the "Brundtland Report" greatly influenced the terms of debate and eventually the trajectory of UN environmental governance, its practical effect in 1987 was modest. This chapter explains why institutional change did not occur in the months following the report's release. A detailed investigation of archival materials reveals that, though prominent cooperative possibilities existed in 1987, states faced coordination problems that resulted in a temporal divergence of expectations that hindered the realization of beneficial institutional change.

This chapter provides background on the origins of the commission as a means of informing the analysis that follows. I then assess the state of the world environment and UN environmental institutions in the mid-1980s, underlining the crucial changes in the international problem structure that underpinned the commission's observations.

[1] WCED 1987.

Although shared international problems presented incentives that would form the basis for agreement at the 1992 Earth Summit, states faced important coordination problems in 1987 that hampered negotiations. The absence of a discrete and unique time frame that could enable a temporal convergence of expectations among actors contributed to negotiations that failed to capitalize on the availability of incentives to alter institutional arrangements. The chapter concludes with an assessment of the theoretical implications of the Brundtland case for this study.

The Origins of the Brundtland Commission

Concerned by the loss of momentum in global environmental governance, the Canadian government proposed the creation of an independent commission at the ninth UNEP Governing Council session.[2] Canada had hoped to negotiate the proposed commission's terms of reference and mandate in 1981 so that it could be formally launched at the 1982 UN "Session of a Special Character" (UNSSC).[3] Although the Canadian proposal encountered resistance from some quarters, it was picked up and spearheaded in a rather clumsy fashion by the Japanese delegation to the 1982 session. The Japanese proposal, recalled one participant, "went over like a lead balloon."[4] While the 1982 session debated the virtues of a commission and formally recommended its creation, efforts were primarily directed toward a face-saving outcome.[5] It was clear that no political basis existed for launching an independent commission in 1982. It took a year of tortuous and divisive negotiations to build the necessary support within the UN General Assembly for the establishment of a special commission in December 1983 to develop an "environmental perspective to the year 2000 and beyond."[6]

The commission concept was highly contentious and was opposed bitterly by UNEP's Executive Director, Dr. Mostafa Tolba.[7] Some argued that the act of launching a commission would be tantamount

[2] United Nations Environment Programme 1981. [3] Bruce 1991, 8.

[4] Dr. Martin Holdgate Interview with the author, August 6, 2012.

[5] For the resolution, see: United Nations Environment Programme 1982.

[6] United Nations General Assembly 1983.

[7] Tolba made a special trip to Japan to try to get the initiative stopped. Lars Björkbom Interview with the author, July 24, 2012.

to admitting the failure of existing institutional arrangements; consequently, UNEP sought to protect its turf by opposing the commission's creation.[8] UNEP had recently aimed to establish its competence for environmental assessment and review through its publication of *The World Environment 1972–1982*.[9] By extolling the virtues of an independent perspective, the commission report seemed to undermine UNEP's goal. Leading developed states, especially the United States and United Kingdom, feared that the commission was likely to condemn their growth-oriented economic policies and advance recommendations that they would be compelled to reject with high political costs.[10] The Soviet Union was strongly opposed to the commission idea and attempted to block it.[11] Even proponents of the commission were not sanguine about its prospects for success. The disappointing returns on the reports of the Independent Commission on International Development Issues in 1980, under former West German Chancellor Willy Brandt, and the Independent Commission on Disarmament and Security in 1982, under former Swedish Prime Minister Olof Palme, did not augur well for the environmental commission.[12]

The UN General Assembly resolution that formally authorized the commission was of no comfort to proponents. The commission's secretary-general, Dr. Jim MacNeill, saw the General Assembly resolution 38/161 as a potential "hydra that would wrap the commission in knots and guarantee failure," because it

> seemed most intent on protecting the turf of UNEP, the key sponsoring agency, and typical of many UN resolutions, it was full of compromises and contradictions that lent themselves to an endless variety of interpretations. I felt that it was an unintended trap and feared that, even if the commission got off the ground, with that resolution as its mandate, it could never be independent and could well become paralyzed.[13]

In addition to an ambiguous mandate, the fact that the new body was referred to as a "special commission" by the General Assembly, rather

[8] Dr. Jim MacNeill Interview with the author, July 2, 2013.
[9] Holdgate et al. 1982. [10] MacNeill 2007, 242.
[11] Dr. Martin Holdgate Interview with the author, August 6, 2012.
[12] Independent Commission on International Development Issues 1982.
[13] MacNeill 2007, 242–243. Brundtland herself shared this appreciation. See: Brundtland 2005, 195.

than an "independent commission," as originally proposed, indicated that it was not meant to be especially enterprising or autonomous. The launch of a parallel – and redundant – Intergovernmental Intersessional Preparatory Committee (IIPC), under UNEP auspices, that would also provide an environmental perspective to the year 2000 and beyond reflected the apprehensions of UNEP and its supporters over the Brundtland Commission.

Shortly after the commission concept was approved, intensive negotiations occurred over its leadership. General Assembly resolution 38/161 called on the UN secretary-general to appoint the commission's chair and vice chair.[14] Although UN Secretary-General, Javier Pérez de Cuéllar, initially favored former Senegalese President Leopold Senghor to be the chair, Tolba opposed his candidacy vigorously.[15] Cuéllar eventually appointed the once and future Prime Minister of Norway, Gro Harlem Brundtland.[16] Tolba successfully insisted on the appointment of Mansour Khalid, Sudan's former foreign minister, as vice chair.

Through Khalid, Tolba attempted to exercise control over the commission.[17] The vice chair took to telexing confidential WCED materials to UNEP's executive director and would regularly visit the commission's office in Geneva, collecting draft papers to surreptitiously send to Nairobi.[18] Tolba argued that the commission should be legally constituted as a United Nations body, an arrangement that would constrain its operations. UNEP sought to exercise control over WCED's finances, banking, staffing, and contracts.[19] Tolba also endeavored to undermine the commission's political basis, especially

[14] UNGAR 38/161 1983.

[15] Other candidates for the chair of the commission were former US President Jimmy Carter and Prince Charles. Lars Björkbom Interview with the author, July 24, 2012. Charles Caccia to Allan J. MacEachen, December 2, 1983, Strong Collection – ESPP, Box 311, Folder 2953.

[16] Brundtland 2005, 194.

[17] On Khalid, Brundtland wrote in her memoirs: "Mansour Khalid considered me a relatively young and inexperienced woman and clearly believed that it would be easy to wrest control from me. He would be disappointed." See: Brundtland 2005, 194.

[18] Dr. Jim MacNeill Interview with the author, July 2, 2013.

[19] Brundtland 2005, 194; MacNeill 2007, 243.

among African countries, by casting aspersions on the political ends of the commission's sponsoring countries.[20]

Archival sources show that Brundtland and Khalid clashed repeatedly over the commission's composition, staffing, and mandate. At the root of their dispute were North-South politics. In his private communications with the WCED chair, Khalid complained vehemently that the commission was "becoming an increasingly western affair, run from the extreme North."[21] In the appointment of the other commissioners, the joint responsibility of the chair and vice chair, Khalid stressed the need for a strong developing country majority among commission members. Although there was debate over appropriate representational precedents, Brundtland endorsed the idea of having a developing country majority.[22] Twelve commissioners came from the developing world, seven from developed countries, and four from Communist states. Serving on a personal basis, the twenty-three commissioners came from all of the world's regions and major states.[23]

Brundtland and Khalid disagreed on the role of the commission's secretariat. Brundtland insisted that Khalid had no authority over secretariat appointments, and proceeded without consulting him when she approached Jim MacNeill, then Director of Environment at the Organization for Economic Co-operation and Development (OECD), about a role as WCED's secretary-general and ex officio member of the

20 CANEMB-Nairobi (David Miller) to DEA-ETE, "Environment Commission," March 22, 1984, Strong Collection, Harvard University Environmental Science and Public Policy Archives (ESPP), Box 311, Folder 2953.

21 Mansour Khalid to Gro Harlem Brundtland, April 3, 1984, Strong Collection – ESPP, Box 311, Folder 2954.

22 Mansour Khalid to Gro Harlem Brundtland, April 3, 1984, Strong Collection – ESPP, Box 311, Folder 2954.

23 In addition to Brundtland and Khalid, the commission was composed of: Susanna Agnelli (Italy), Saleh Abdulrahman Al-Athel (Saudi Arabia), Pablo Gonzalez Casanova (Mexico), Bernard T. G. Chidzero (Zimbabwe), Volker Hauff (West German), Istvan Lang (Hungary), Ma Shijun (China), Margarita Marino de Botero (Colombia), Nagendra Singh (India), Paulo Nogueira-Neto (Brazil), Saburo Okita (Japan), Shidath S. Ramphal (Guyana), William Doyle Ruckelshaus (United States), Mohamed Sahnoun (Algeria), Emil Salim (Indonesia), Bukar Shaib (Nigeria), Vladimir Sokolov (Soviet Union), Janez Stanovnik (Yugoslavia), Maurice Strong (Canada), and Jim MacNeill (Canada). It should be noted that, as WCED Secretary-General, MacNeill was an ex officio member of the commission. Pablo Gonzalez Casanova of Mexico would leave the commission in August 1986, leaving the commission with twenty-two members when *Our Common Future* was released.

commission.[24] Khalid decried MacNeill's appointment, arguing that his background with the OECD was a "major liability in the eyes of the South." To think that the MacNeill appointment would pass unnoticed in developing countries, he maintained, "is simply flying in the face of reality." Khalid opposed MacNeill's title of secretary-general, arguing that it "implies a certain measure of political independence not only in the exercise of judgment but in initiating, which is certainly not envisaged here."[25] In a subsequent letter to all the commissioners, Khalid protested Brundtland's actions, contending that a strong secretariat would undermine the authority of commissioners.[26] The commissioners, not the secretariat, should be in the political and intellectual lead.

A final area of conflict between WCED's chair and vice chair concerned their respective responsibilities. Khalid argued that he should have an effective veto over most substantive commission decisions so as to "achieve the collective leadership envisaged by [the] G.A. Resolution at the level of the Chairman and Vice Chairman, representing the North and the South respectively."[27] Because of the well-documented misgivings of countries concerning the commission, he argued, it should not act in an especially entrepreneurial fashion. International political sensitivities distinguished the environmental commission from the relatively autonomous Brandt or Palme Commissions. A "Brundtland Commission" was certainly not in the cards, Khalid jabbed.[28] Brundtland, for her part, noted that the General Assembly had not appointed co-chairs and made it clear that she did not envision a "twin headed Commission or any other kind of collective leadership."[29]

24 Gro Harlem Brundtland to Mansour Khalid, May 11, 1984, Strong Collection – ESPP, Box 311, Folder 2953.
25 Mansour Khalid to Gro Harlem Brundtland, April 3, 1984, Strong Collection – ESPP, Box 311, Folder 2954.
26 Mansour Khalid to the Members of the World Commission on Environment and Development, May 11, 1984, Strong Collection – ESPP, Box 311, Folder 2953.
27 Mansour Khalid to Gro Harlem Brundtland, March 1984, Strong Collection – ESPP, Box 311, Folder 2954.
28 Mansour Khalid to Gro Harlem Brundtland, April 3, 1984, Strong Collection – ESPP, Box 311, Folder 2954.
29 Gro Harlem Brundtland to Mansour Khalid, April 30, 1984, Strong Collection – ESPP, Box 311, Folder 2953.

The dispute went all the way to the UN secretary-general's office, where Brundtland's interpretation of General Assembly Resolution 38/161 prevailed.[30] The political backing of numerous sponsoring countries helped her case. Some governments, for example, threatened to reduce their funding of UNEP if Tolba and Khalid did not back off in their attempts to control the commission and its findings.[31] The Canadian government made its substantial financial contribution to the commission conditional on WCED's political independence of UNEP.[32]

Though there were ongoing frustrations, especially in WCED's relations with the IIPC, Brundtland had seized political control of the commission by summer 1984. From here, recalled Brundtland in an interview with the author, "Tolba's influence receded gradually."[33] From 1984 until early 1987, the commission held public hearings and site visits in eight countries.[34] The public hearings, unprecedented to that point and seen by some as an invasion of sovereignty, raised the profile of the commission and provided an important information channel for the Brundtland report.[35] In addition to the hearings, WCED sponsored 75 studies and received more than 800 papers and reports that formed the basis for *Our Common Future*.[36] The conflict over control of the commission in 1983 and early 1984 was, however, just a manifestation of the fractious world political order to which the commission would report.

30 Dr. Jim MacNeill Interview with the author, July 2, 2013. Bob Munro Interview with the author, July 3, 2013.

31 Lars Björkbom Interview with the author, July 24, 2012.

32 Geoffrey Bruce to Maurice Strong, "International Commission on the Environment – To the Year 2000," October 14, 1983, Strong Collection – ESPP, Box 311, Folder 2953.

33 Gro Harlem Brundtland Interview with the author, August 17, 2021.

34 The nine commission meetings were held in: Geneva, Switzerland, October 1–3, 1984; Jakarta, Indonesia, March 27–31, 1985; Oslo, Norway, June 21–28, 1985; São Paulo, Brazil, October 25–November 4, 1985; Ottawa, Canada, May 21–31, 1986; Harare, Zimbabwe, September 15–19, 1986; Nairobi, Kenya, September 20–23, 1986; Moscow, USSR, December 6–12, 1986; and Tokyo, Japan, February 23–28, 1987.

35 MacNeill 2007, 244. Bob Munro Interview with the author, July 3, 2013.

36 Munro 1991, 14.

Problem Setting

In the mid-1980s, key environmental indicators suggested that pollution was worsening and that natural resource depletion was continuing apace. These findings were punctuated by a string of human and environmental disasters that demonstrated dramatically the consequences of ongoing environmental neglect.[37] The Third World debt crisis, which reached dire proportions in the summer of 1982, greatly exacerbated environmental decline in developing lands through the 1980s. These problems rendered existing institutional arrangements deficient in important respects.

Exogenous Change

As WCED was deliberating, signs of global environmental deterioration were ubiquitous. Climate change was an ominous challenge, especially as global energy consumption was expected to climb in the coming decades.[38] In 1986, scientists detected a "hole" in the ozone layer over Antarctica, confirming for many the scale of human impacts on planetary systems.[39] Acid rain, caused by emissions of sulfur dioxide and nitrous oxide, was a serious problem throughout the world. It was estimated that acid precipitation was costing the seventeen states in the eastern United States US$7 billion annually in material corrosion and reducing annual Japanese wheat and rice crop production by as much as 30 percent.[40]

Signs of environmental deterioration were increasingly present in the global South. At WCED's fourth meeting, held in São Paulo, Brazil, in autumn 1985, commissioners acknowledged that, in the last decade, "the epicenter of environmental problems ha[d] moved from North to South and that many of those problems, particularly in the South were

[37] There was an interesting development in thinking in the mid-1980s about the impact of human institutions on states' capacity to respond to environmental disasters. The effects of drought, for example, could be reduced greatly by effective social institutions. This thinking was influential and germane to the findings of the Brundtland Commission. For an interesting perspective from this period, see: Wijkman 1984.

[38] WCED 1987, 170–174.

[39] National Science Foundation 1986.

[40] WCED 1987, 178–181.

related to poverty."[41] Chinese industry, which relied heavily on coal and outdated furnaces for power, caused severe air pollution in urban settlements. Kuala Lumpur had two to three times the air pollution of major US cities. Of India's 3,119 towns and cities, only 8 had full sewage treatment facilities.[42] In addition to the millions likely to experience increasingly frequent drought, 850 million people were vulnerable to desertification.[43] The high cost of energy increased pressure on tropical forests, leading to widespread timber-felling.[44]

The environmental problems of the developing world were likely to become an increasing focus in the twenty-first century for at least three reasons. First, while the resources necessary to ameliorate environmental damage were available in the industrialized states, air and water pollution would grow worse in developing countries because they lacked the resources and technical capacity to manage them effectively. Second, as an internal WCED discussion paper noted, there had been a marked change in the "industrial geography of the world, with the locus of development of resource-intensive and energy intensive industries shifting to the South."[45] This changing reality had implications for the character of global resource management and pollution.

Third, and perhaps more importantly, the pace of environmental deterioration in the South was quickly increasing because of the Third World debt crisis. The oil shocks of the 1970s brought an influx of funds from oil-rich countries to commercial banks, facilitating low-interest lending to developing countries. Cheap money led many developing countries to fuel economic growth through heavy borrowing. Between 1975 and 1983, there was a quadrupling of

41 "Minutes of the Fourth Meeting of the World Commission on Environment and Development, October 25–November 4, 1985, São Paulo, Brazil," Strong Collection – ESPP, Box 315, Folder 2889, 14.

42 WCED 1987, 240.

43 Dr. Mostafa Tolba, "The Premises of Sustainable Development, Address by Dr. Mostafa Tolba, Executive Director of the United Nations Environment Programme to the Special Commission on Environment and Development, Geneva, October 1, 1984," Strong Collection – ESPP, Box 313, Folder 2967.

44 WCED Discussion Paper, "International Economic Relations, Environment and Development," WCED/85/23, Strong Collection – ESPP, Box 315, Folder 2889.

45 "International Economic Relations, Environment and Development," in: Minutes of the Fourth Meeting of the World Commission on Environment and Development, October 25–November 4, 1985, São Paulo, Brazil, Strong Collection – ESPP, Box 315, Folder 2889, 14.

foreign debt in Latin America, from US$75 billion to US$315 billion.[46] The situation was similar in Africa. By 1988, the debt of African countries had climbed to nearly US$140 billion. The debt-to-export ratio among African states quadrupled between 1980 and 1988, reaching 361.2.[47]

Several factors contributed to making this level of debt unsustainable in the 1980s. Falling commodity prices hit debtor countries hard, diminishing export earnings.[48] This was a major challenge because sixty-four developing countries relied on primary commodities for over 50 percent of their exports.[49] A tightening of monetary policy by the United States and some European countries led to higher interest rates, contributing to a US$21 billion increase in annual debt service payments from 1980 to 1987.[50] Developing world access to private lending was curtailed greatly following a Mexican default in August 1982.[51] These factors contributed to a dramatic shift in the world economic balance. In 1980, there was a US$38.1 billion inflow of aggregate net resource transfers to the developing countries.[52] By 1987, this had turned into a US$16.7 billion outflow from the developing to the developed world.[53]

This rapid US$54 billion shift in aggregate net resource transfers had profound implications for UN environmental affairs. The dismal economic condition of much of the developing world took its toll on public finances. Debt service payments, loss of export markets, and a stagnating flow of official development assistance limited the ability of developing countries to devote resources to environmental protection. Enforcement and implementation of environmental regulations

[46] Institute of Latin American Studies 1986, 69.

[47] Stein and Wayne Nafziger 1991, 174.

[48] Between 1980 and 1984 falling commodity prices led to a loss of about $55 billion in export earnings for developing countries. See: WCED 1987, 36.

[49] "International Economic Relations, Environment and Development," in: Minutes of the Fourth Meeting of the World Commission on Environment and Development, October 25–November 4, 1985, São Paulo, Brazil, Strong Collection – ESPP, Box 315, Folder 2889.

[50] Culpepper 1991, 5.

[51] On Mexican macroeconomic policy and prospects at the time, see: Rojas-Suarez 1992.

[52] Aggregate net resource transfers are calculated by finding the difference between the net movement of official and private capital and the net movement of interest, dividends, and profits.

[53] Culpepper 1991, 5.

suffered badly and the budgets of environment departments were in retreat. The need to service foreign debts with export earnings contributed to short-term pressures to exploit natural resources, thus accelerating the rate of desertification, soil erosion, and salinization.[54] In many countries, cash crops pushed food crop production to marginal lands, thereby degrading soils and deepening poverty.[55] Tropical forests were being cut down rapidly to make space for export-oriented beef production.[56]

In addition to scientific indicators, several prominent human and environmental disasters in the mid-1980s brought public attention to the worsened state of the world environment.[57] Severe drought caused a devastating famine from 1983 to 1985 in Ethiopia, which resulted in more than 400,000 deaths. In November 1984, a liquid gas storage tank exploded outside Mexico City, killing 1,000 people and leaving thousands more homeless. In December 1984, an industrial gas leak in a pesticide plant in Bhopal, India, exposed more than 500,000 people to methyl isocyanate.[58] In April 1986, a fire and explosion at a nuclear power plant in Chernobyl, Ukraine, released radioactive particles, which spread across much of Europe. A November 1986 agrochemical leak in Basel, Switzerland, released toxic chemicals into the air and water, temporarily turning sections of the Rhine River red and devastating local wildlife.[59] These high-profile events, among others, in the

[54] WCED Discussion Paper, "International Economic Relations, Environment and Development," WCED/85/23, Strong Collection – ESPP, Box 315, Folder 2889.

[55] WCED 1987, 127–128.

[56] "International Economic Relations, Environment and Development," in: Minutes of the Fourth Meeting of the World Commission on Environment and Development, October 25–November 4, 1985, São Paulo, Brazil, Strong Collection – ESPP, Box 315, Folder 2889.

[57] For a discussion of the public and political effect of these events, see: Shabecoff 1996, 94–95.

[58] The Bhopal disaster had a particularly important effect on WCED because it happened relatively early in the commission's deliberations and seemed to underline many of the lessons being drawn from the 1983 to 1985 Ethiopian famine. At the commission's second meeting, held in March 1985, an extensive discussion of the legal and institutional implications of the Bhopal incident was held. Discussions underscored the need for international legal and institutional capacity building. See: "Minutes of the Second Meetings, 25–31 May, Jakarta, Indonesia," Strong Collection – ESPP, Box 317, Folder 3010.

[59] For WCED's discussion of these events, see: WCED 1987, 228–299.

1980s, highlighted the limitations of the current "react and retrofit" approach to global environmental management.[60]

When *Our Common Future* was released in 1987, fifteen years after the Stockholm conference, evidence of a significant worsening of global environmental problems was abundant. Scientific research and other indicators suggested that global problems, such as climate change, ozone depletion, and acid rain, were worsening progressively. Unlike at Stockholm, environmental problems were increasingly centered in the developing world. Poverty, as much as affluence, was an important source of environmental troubles such as desertification, tropical deforestation, and poor sanitation. The Third World debt crisis badly exacerbated these issues, accelerating environmental decline.[61]

Institutional Deficiencies

The changing international problem-setting called into question the efficacy of existing institutional arrangements. Two main defects were important. First, UNEP's inability to serve as a clearinghouse for environmental policy coordination diminished the effectiveness of the UN environmental enterprise, often contributing to outright contradictions in the policies of UN specialized agencies. Second, the treatment of environment as another "sectoral" subject in the UN system was the root of poor coordination, and ensured that environment and development ends were not managed as an integral whole. Consequently, UN environmental policy was fundamentally reactionary.

In 1972, states had layered UNEP on top of existing institutional measures, giving rise to certain difficulties. Although it was mandated to serve as a focal point for catalysis and coordination within the UN system, UNEP lacked the authority and resources to fulfill its charge.[62]

60 WCED 1987, 7.

61 Though it would reach more modest conclusions than WCED, the intergovernmental IIPC also concurred. In November 1986, the committee noted that, given the changing problem structure, current institutional arrangements "may not be adequate anymore." See: Intergovernmental Inter-Sessional Preparatory Committee 1986.

62 UNEP also lacked the resources to induce the cooperation of the specialized agencies. From 1972 to 1987, the UNEP Environment Fund had received a total contribution of US $400 million, less than half of the *annual* budgets of its main partner agencies – the World Health Organization, the Food and Agriculture

The System-Wide Medium Term Environment Programme – UNEP's latest attempt to achieve its coordination mandate – was hindered by unpredictable funding and inconsistencies in governmental policy positions in different UN intergovernmental bodies.[63] The result, wrote the former deputy executive director of UNEP, was that "under existing conditions the UN system's influence [on environmental affairs] is often fragmented and less effective than it might be because of the independent character of the specialized agencies and endemic weakness of co-ordination."[64] Institutions, an internal WCED discussion paper remarked, "have a congenial compulsion to classify problems in their own image." Despite numerous environmental conventions and agreements, the implementation of environmental measures lacked coherence.[65]

The root of the coordination quandary was that environment was institutionalized as a sectoral issue, rather than an integrative one that stretched across multiple realms of institutional activity. At the international level, there was an obvious inconsistency between the global environmental order and international policies on development lending, trade and investment, and agricultural development.[66] Environment was thus treated separately from every other element of economic activity. The environmental policies of UNEP often conflicted with the developmental aims of the UN Development Programme (UNDP) or the Food and Agriculture Organization (FAO), for example.[67] The world community was simultaneously setting ecological fires with one hand and attempting to extinguishing

Organization, and the United Nations Educational, Scientific, and Cultural Organization. Munro 1991, 18.

63 On coordination, a note by UNEP's executive director observed: "Often the Governing Council is not always clear when it is acting as Governing Council for the UN system environmental activities as a whole and when it is guiding the UNEP secretariat. Further, it was anomalous that Governments clamored for coordination when they did not achieve it between their own delegations to governing bodies of different UN agencies." See: United Nations Environment Programme 1986.

64 Thacher 1991, 49.

65 "Revised Outline for Discussion – Chapter 12: International Co-operation, Peace and Sustainable Development," July 14, 1986, Strong Collection – ESPP, Box 315, Folder 2993.

66 WCED 1987, 10.

67 UNEP efforts to limit the use of pesticides for environmental reasons was stymied by the UN's Food and Agriculture Organization, which was, at the same time, promoting a fivefold increase in the use of pesticides. See: Goldsmith 1982.

them with the other. The inability of existing institutional arrangements to take wider economic factors into account led states to conclude that UNEP could not adequately address the *sources* of global environmental deterioration.[68]

Nor was there an integrated UN environment and development agenda. Environmental policies took development as a given. Developmental policies did not, except on the margins, take the environment into account.[69] The current setup made greater sense in the late 1960s or early 1970s, when environmental problems were more concentrated in industrialized countries and were aimed primarily at repairing environmental damage that had already occurred.[70] Current institutions were not well-suited to managing the developmental dilemmas of the global South or preventing damage in the first place. Surveying the changing nature of environmental questions, an internal WCED discussion paper noted: "The fact is that the swift evolution of objective realities has now outdistanced most of our institutions to such an extent that, without change, they cannot succeed." The paper later stated: "The difficult fact is that, as presently conceived and organized, most of our existing institutions are by mandate and structure incapable of handling the new class of issues."[71]

The Brundtland Report

Environmental issues were institutionalized within the United Nations system as a sectoral, rather than a horizontal, integrative issue. The changing character of planetary decline made a react-and-retrofit approach to global environmental management increasingly unsuitable. Though dozens of other sectoral agencies, such as the WHO and FAO, dealt with issues that had deep environmental implications, they often did not take environmental considerations into account. UNEP, small and poorly funded, lacked the means to coordinate

[68] Lars Björkbom Interview with the author, July 24, 2012.

[69] "Statement to the Intergovernmental Inter-sessional Preparatory Committee by Mr. Jim MacNeill, Secretary General, World Commission on Environment and Development, Nairobi, 7 March 1986," Strong Collection – ESPP, Box 318, Folder 3014.

[70] WCED 1987, 9–10.

[71] "Revised Outline for Discussion, Chapter 12, International Co-operation, Peace and Sustainable Development," May 1986, Strong Collection – ESPP, Box 315, Folder 2993.

larger, better established UN organs. This contributed to, rather than combated, mounting environmental deterioration on a global scale. As countries, especially those of the developing world, endeavored to accelerate their development processes – often with inadequate institutional and technical capacity – the shortcomings of UN environmental institutions manifested themselves in a string of human and environmental disasters.

It was in this context that WCED responded to the call of the UN General Assembly to provide an environmental perspective to the year 2000 and beyond. At the inaugural commission meeting from October 1 to 3, 1984, the WCED mandate was framed in terms of a choice between the "standard" environmental agenda and an "alternative" agenda. The standard agenda was primarily concerned with environmental pollution, human settlements, and natural resource management. If the commission followed the standard approach, its recommendations would have aimed chiefly to provide suggestions on how states could better achieve the goals that they had been targeting since Stockholm. An alternative agenda would have an economic and trade policy focus, concentrating on the *sources* of environmental problems. It would have a much broader focus on the UN system as a whole, and would seek to move from a "react-and-cure" to an "anticipate-and-prevent" approach.[72]

The alternative agenda had at least four advantages compared to the standard agenda. First, by explicitly integrating economic development imperatives with environment, the alternative agenda was more directly relevant to developing countries than the standard agenda. Since most countries in the world – and in the UN General Assembly, where the WCED report would be presented – were developing economically, the recommendations became relevant to a wider subsection of states. Second, by addressing itself to the policies of economic and social decision-makers, the alternative agenda reached a wider, more influential governmental audience. Third, the alternative agenda opened the way to new possibilities for preventive, cost-effective actions that could utilize a wider array of economic and regulatory policy instruments. Fourth, the new agenda provided a rationale for a reconsideration of

[72] WCED Secretariat, "Mandate: Key Issues, Strategy and Work Plan of the World Commission on Environment and Development," October 22, 1984, Strong Collection – ESPP, Box 313, Folder 2958, 19, 24, 29–32.

institutional questions. In view of the limitations and institutional weakness of UNEP, there was a growing need to reassess fundamentally institutional arrangements.[73]

The principal proponent and author of the alternative agenda was Jim MacNeill, WCED's secretary-general. Drawing heavily on his experience as Environment Director at the OECD, MacNeill wrote a draft mandate document and secured Brundtland's assurance of support for the alternative agenda before agreeing to head the commission's secretariat.[74] With Brundtland's backing, the alternative agenda was endorsed by commissioners as the basis for the commission's work at WCED's inaugural meeting in October 1984.[75] The MacNeill-drafted mandate document, titled "Mandate for Change," was widely circulated and welcomed by hundreds of governmental and nongovernmental actors.

The alternative agenda was a sustainable development one. Although the concept of sustainable development was first used in the 1980 World Conservation Strategy, WCED popularized it and provided its definitive definition: development that "meets the needs of the present without compromising the ability of future generations to meet their own needs."[76] WCED argued that sustainable development did not imply absolute limits to economic growth. The only limits imposed on development should be based on current technological capacities and on the ability of the biosphere to absorb the effects of human activities.[77] If intergenerational equity was to be realized, no one generation could irretrievably damage the earth's systems.[78]

[73] This discussion is derived from: Munro 1991, 12–13. WCED Secretariat, "Mandate: Key Issues, Strategy and Work Plan of the World Commission on Environment and Development," October 22, 1984, Strong Collection – ESPP, Box 313, Folder 2958.

[74] Dr. Jim MacNeill Interview with the author, July 2, 2013. As OECD Environment Director, MacNeill had organized an influential conference on "Environment and Economics" in June 1984. The results were published in: OECD 1985.

[75] WCED, "Minutes of the Inaugural Meeting, 1–3 October 1984," Strong Collection – ESPP, Box 312, Folder 2962.

[76] IUCN 1980; WCED 1987, 8.

[77] WCED 1987, 8, 31–32.

[78] Steven Bernstein describes the contribution of the Brundtland report as an essentially Keynesian approach to the environment–development relationship. It cemented the view that growth and environmental protection are compatible within a framework of "managed sustainable growth." See: Bernstein 2001, 6, 52.

To advance sustainable development, the commission advocated sweeping institutional change. In view of the strong institutionalized opposition faced by WCED, the unusual "hardheadedness" of the commission's institutional recommendations was a notable feature of the Brundtland report.[79] During its preparation, MacNeill noted to the author, he guarded the contents of chapter 12 on institutional and legal change especially closely. Drafts of the chapter were not made available to commissioners until late 1986 – just months before the commission's report was scheduled to be released – so as to minimize the anticipated pushback from UNEP and from national environment ministries.[80]

The commission argued that the United Nations secretary-general should exercise primary responsibility for system-wide interagency coordination and cooperation, thus providing a high-level center for leadership. This high-level center should, *Our Common Future* argued, assess, advise, assist, and report on progress made toward – and needed for – the implementation of sustainable development within the UN system. Governments should make arrangements at the General Assembly and on the intergovernmental bodies of UN specialized agencies to reinforce the secretary-general's coordinative authority.[81] The secretary-general should, furthermore, establish a UN Board for Sustainable Development. The board would combat interagency fragmentation by agreeing on combined tasks to be executed on an interagency basis.[82]

Within governments, central decision-making bodies should play a more active role in the implementation of sustainable development, ensuring greater accountability for economic development policies.[83] This would necessarily include participation from central economic, financial, trade, energy, and agricultural ministries, for example.[84] At the international level, the World Bank, International Monetary Fund,

[79] Brenton 1994, 128.

[80] Because so many commission drafts had found their way to Mostafa Tolba, usually via Mansour Khalid, MacNeill composed chapter 12 of *Our Common Future* at his home in Geneva. For an extra measure of security, most early drafts were prepared longhand. Dr. Jim MacNeill Interview with the author, July 2, 2013.

[81] WCED 1987, 318. [82] WCED 1987, 318–319. [83] WCED 1987, 314.

[84] "Revised Outline for Discussion – Chapter 12: 'International Co-operation, Peace and Sustainable Development,'" July 14, 1986, Strong Collection – ESPP, Box 315, Folder 2993.

and the regional development banks were crucial actors. These institutions had often been criticized for funding short-sighted development projects that destroyed forests, damaged watersheds, and undermined natural systems.[85] The Brundtland report singled out the World Bank as especially important in taking a lead to ensure that its lending policies advanced sustainable development. To accomplish this end, the bank's internal structure and processes would need to be transformed and reoriented.[86]

The institutionalization of sustainable development across the UN system had to be accompanied, argued Brundtland, with "substantially increased aid flows," measures to reduce Third World debt, and development-friendly adjustments to North-South terms of trade.[87] A "new deal in international cooperation" would allow developing countries to build the institutional capacity necessary to grow sustainably.[88] After contending that assessed or voluntary financial contributions would not provide a sufficient basis for implementing sustainable development internationally, the Brundtland report argued that the General Assembly should give careful consideration to automatic funding options. In particular, revenue could be generated from taxing the use of the international commons and natural resources.[89]

The commission urged states to negotiate a universal declaration and convention on environmental protection and sustainable development. UN environmental institutions had, the commission stressed, reached the point where some means of consolidating and extending existing legal principles was required, lest the system become incoherent. The 1972 Stockholm declaration and the 1982 Nairobi declaration, as well as the many international conventions that had been negotiated, would form the basis for this initiative.[90] Drawing on WCED's expert panel of international legal experts, *Our Common Future* submitted a set of twenty-two articles to the international community as an annex to its report.[91]

Within this broader agenda, WCED argued that UNEP should play a major part in environmental protection issues. Accordingly, UNEP's

[85] Shabecoff 1987b, A14. [86] WCED 1987, 337–338.
[87] Brundtland 1987a. See also: Brundtland 1987b.
[88] For a detailed treatment of the economic ideas that underpinned the commission's findings, see: Bernstein 2001, 61–69.
[89] WCED 1987, 340–342. [90] WCED 1987, 332–333.
[91] WCED 1987, 348–351.

programmatic work should be refocused to enhance the availability of environmental data, assessment, and reporting, and to provide support for environmental management. To enable it to play a leadership role in environmental protection in the UN system, UNEP's catalytic and coordination role should be reinforced and its Environment Fund enlarged. UNEP would continue to promote scientific and technological research, as well as to develop criteria and indicators for environmental quality standards. To assist the World Bank, UNDP, and other development organizations, UNEP should provide advice on the environmental aspects of development programs.[92]

The ambitious proposals for institutional and legal change contained in *Our Common Future* provided a detailed basis for a complete overhaul of UN environment institutions, focusing on the economic drivers of environmental decline. By advancing this alternative agenda the Brundland report proposed to tackle the lack of coordination in the UN system by relocating the main responsibility to the UN secretary-general and a Board for Sustainable Development. The new conceptual apparatus of sustainable development moved environment to the center of the UN system, addressing the ineffective sectoral treatment of existing institutional measures. Environment would, if WCED's recommendations were adopted, be managed as a horizontal, integral issue. To review progress on the implementation of sustainable development, WCED recommended a series of conferences on a local, global, and international basis.

A Temporal Focal Point?

The commission's report articulated an urgent and compelling rationale for sweeping change in UN environmental institutions. Did this clarion call constitute a Temporal Focal Point within the UN system? The short answer is no. This section shows that the publication of the Brundtland report cannot be considered a Temporal Focal Point in the history of UN environmental politics. Although the consideration of *Our Common Future* at the UN General Assembly in autumn 1987 was both a unique and highly conspicuous moment along the temporal continuum, no definite negotiation timeline for implementing its recommendations emerged in UN bargaining.

[92] WCED 1987, 319–323.

The information strategy of the World Commission on Environment and Development was a subject of great emphasis in the months and years leading up to the report's presentation to the General Assembly in 1987. This was a lesson learned by Brundtland and MacNeill from the underwhelming returns on the Brandt and Palme commissions.[93] The information strategy was multifaceted and had been on the commission agenda since its inaugural meeting.[94] WCED's early planning touched on the openness of its deliberations, the style and content of its report, and its relations with media and nongovernmental actors. An "advocacy phase" following the completion of the report was also planned.[95] The aim of this strategy was to ensure that WCED received the widest possible exposure, preparing public and government opinion for its recommendations so as to facilitate swift adoption.

Although *Our Common Future* benefited from being released at a high point of public concern about the state of the world environment, the commission was in large part responsible for making its findings highly conspicuous.[96] Following WCED's inaugural meeting, its secretariat circulated the commission's mandate document to almost 1,000 ministers and senior officials, heads of governmental and nongovernmental organizations, scientific bodies, citizens' groups, and media. This "unprecedented open posture" was unique in the history of commissions and exposed many key actors to the alternative, sustainable development agenda.[97]

WCED's public hearings were a hallmark of its work, connecting actors from many of the world's regions to the commission and its mandate. Public hearings were held in Jakarta, Oslo, São Paulo, Ottawa, Harare, Nairobi, Moscow, and Tokyo. The concept of public

93 Brundtland 2005, 195; MacNeill 2007, 244.

94 "Draft Information Strategy: Note by the Secretary General," WCED/85/11, March 14, 1985, Strong Collection – ESPP, Box 316, Folder 2994.

95 On the commission's work, Bob Munro recalled: "No other international body or world commission has ever used such innovative approaches as joint government/industry/NGO/expert groups linked directly to the decision-making process, the public release of preliminary findings and reports for comment or the convening of open public hearings for getting and testing new ideas and approaches." See: Munro 1991, 14.

96 Public concern had been aroused by the environmental disasters discussed above. For a discussion of the waves in public concern about environment, see: Brenton 1994, 55–56.

97 "Draft Information Strategy: Note by the Secretary General," WCED/85/11, March 14, 1985, Strong Collection – ESPP, Box 316, Folder 2994.

hearings had been highly controversial and many argued, not incorrectly, that, based on General Assembly resolution 38/161, the commission had no authority to conduct them. Some countries, including Indonesia, harbored deep reservations about the commission's insistence on setting of the agenda for hearings, inviting whomever it desired, and opening the proceedings to the media.[98] The hearings became a vital, and deliberate, means of raising the commission's public profile, helping WCED build supportive networks within government administrations and among subnational actors. The hearings, remembered MacNeill, "enabled a huge initial outreach by the Commission, which was crucial … to the ultimate impact of the report."[99] Brundtland echoes this view, arguing that the "wide list of inputs and concern for our future, across the world" was *a*, perhaps *the*, most important ingredient in the broad basis of support for the commission's recommendations.[100]

The style and content of *Our Common Future* was another means of ensuring its relevance. From the outset, the commission and its secretariat devoted much attention to the format and style of the final report. The typical UN approach, which was regarded as "long, inaccessible, and unreadable," was quickly ruled out.[101] The report, it was argued by the WCED secretariat, should engage a wide readership "whose support may be crucial to the overall purposes of the commission," avoiding the tendency of such reports primarily to "satisfy those who have written [them]."[102] In addition to being made available as a UN document, *Our Common Future* was published by Oxford University Press and, less than a year after its release, had sold more than 60,000 copies in English and had been translated into five languages. It was later published in eleven other languages.[103] The popularity of *Our Common Future* boosted the conspicuousness of its recommendations and gave them political weight. Its findings were widely read and discussed. The report became a public phenomenon.

98 Brundtland 2005, 199; MacNeill 2007, 244.
99 MacNeill 2007, 244. For an excellent discussion of the public outreach by the commission, see: Borowy 2014, 157–158.
100 Gro Harlem Brundtland Interview with the author August 17, 2021.
101 "Draft Information Strategy: Note by the Secretary General," WCED/85/11, March 14, 1985, Strong Collection – ESPP, Box 316, Folder 2994.
102 "Format of the Final Report: Note by the Director of Information," WCED/85/20, June 13, 1985, Strong Collection – ESPP, Box 315, Folder 2987.
103 MacNeill 1988, 6.

Vital to the commission's impact was its uniqueness. Although the Intergovernmental Inter-sessional Preparatory Committee also released its environmental perspective in 1987, the profile of *Our Common Future* dwarfed the intergovernmental document.[104] Sympathetic governments insisted that the commission report should be dealt with by the General Assembly in a separate resolution from the IIPC report.[105] In addition, a special General Assembly debate on environment and development was held on October 19, 1987. Efforts to ensure the unique standing of the WCED report were aided greatly by the fact that Gro Harlem Brundtland became prime minister of Norway in May 1986. To promote the report and advance its recommendations, she used the Norwegian foreign ministry and other levers available to a head of government.[106] This gave the WCED report a unique standing at the General Assembly.

Our Common Future's singular status was bolstered by an extensive campaign to raise the public and political awareness of the commission's recommendations. The idea of a nine-month extension to the life of the commission to enable an "advocacy phase" came from Jim MacNeill. Though the decision to extend the commission involved taking further liberties with General Assembly resolution 38/161, commissioners backed the idea at their May 1986 meeting in Ottawa.[107] After the carefully scripted formal launch event for *Our Common Future* on April 27, 1987 at the Queen Elizabeth II Centre in London, commissions traveled extensively throughout the rest of the year to prime world opinion for institutional change. On April 28, Brundtland traveled to Washington to meet with US President Ronald Reagan, congressional leaders, and the president of the World Bank.[108] Concurrently, MacNeill presented the report to the OECD in Paris. In early May, presentations were made to the ministers and commissioners of the European Community, the European Free Trade Agreement counties, and to Comecon members in Hungary. In 1987,

[104] Intergovernmental Inter-sessional Preparatory Committee 1987.

[105] Tolba had initially drafted a single resolution for the approval of the UNEP Governing Council on the IIPC *and* the World Commission reports.

[106] Brundtland 2005, 215–216. [107] MacNeill 1988, 4.

[108] Though Brundtland met with US President Ronald Reagan, it was strictly in her capacity as prime minister of Norway. She was asked not to supply a copy of *Our Common Future* to the US president. Dr. Jim MacNeill Interview with the author, July 2, 2013.

special presentations were also held in Cairo, Nairobi, New Delhi, Beijing, Montevideo, and to the Organization for African Unity in Addis Ababa.[109] From April to October 1987, presidents, prime ministers, or ministers of more than 100 states were briefed on the Brundtland report.[110] These briefings were given great emphasis by the commission because they ensured that senior bureaucrats, as well as political leaders, were familiar with *Our Common Future* and its institutional recommendations.[111]

While the arrival of commission report was both conspicuous and unique, no particular bargaining time frame became salient for the follow-up on the Brundtland Commission's recommendations. The report encouraged regional conferences and, in 1992, an international conference to review progress on instituting sustainable development and promoting effective follow-up.[112] The UN secretary-general supported the concept of regional seminars, including a global conference for youth and educators. Others advocated biennial ministerial meetings preceding sessions of the UNEP Governing Council.[113] While negotiations on a UN General Assembly resolution occurred at the UNEP Governing Council and in the General Assembly's Second Committee in 1987, they did not set in motion a negotiating process geared toward institutionalizing sustainable development across the UN system.

There was a high degree of ambiguity on the timing and form of the negotiation process to put the WCED recommendations into action at the UN. Internal documents show that Gro Harlem Brundtland herself

[109] Details on the commission's travels came from: WCED, "Press Release: 'Our Common Future' – The World Takes Notice," May 6, 1987, Strong Collection – ESPP, Box 392, Folder 3734. Brundtland 2005, 203.

[110] IIED, IUCN, and W. H. Lindner, "Our Common Future: Seizing the Opportunity – A Proposal for Follow-up of the WCED Report 'Our Common Future,'" March 3, 1988, Strong Collection – ESPP, Box 392, Folder 3744.

[111] In an interview, Jim MacNeill described the significance of the briefings in the following manner: "The thing about these meetings with ministers or senior bureaucrats – I knew this from my time in government – was that, if we had a meeting scheduled, the senior bureaucrats would have to read the report, discuss it with their colleagues, and brief the minister. They didn't want to read it, but had to. These meetings were, therefore, critical to our outreach. The leaders and ministers would know what the report said. And, more importantly because political leaders change every three or four years, the senior bureaucrats had read the report." Dr. Jim MacNeill Interview with the author, July 2, 2013.

[112] WCED 1987, 23.

[113] Borowy 2014, 159–160.

had expected as late as mid-October 1987 that the General Assembly could immediately launch a follow-up process on the commission's recommendations.[114] Others on the WCED secretariat estimated that it would take no less than two years to put the commission's recommendations into action.[115] Still others looked toward a still notional international conference in 1992.[116] While these timelines were not necessarily incompatible, they do illustrate that there was a high degree of ambiguity concerning bargaining timelines, even among WCED principals. As will be demonstrated below, ambiguity concerning focal timelines made temporal coordination difficult, even when actors would have benefited from reaching agreement more swiftly.

The International Response to the Brundtland Report

The commission's report addressed important problems in the international system and shed light on institutional incentives that had been opened up by shifts in global conditions. Despite the emergence of new cooperative potentialities, the international response to *Our Common Future* in 1987 was strikingly cautious. This caution did not stem from a rejection of WCED's normative thrust or its reformist orientation.

This section analyzes the international response to the WCED report. First, it summarizes negotiations over the text of what became UN General Assembly resolution 42/187, highlighting the report's chief provisions. The resolution's tepidness, which was highly disappointing to reformers, stands in contrast to the enthusiastic reception that governments gave to *Our Common Future* in the General Assembly. While institutional change in the UN system proceeded haltingly, states moved more definitively in other national and international fora to institutionalize sustainable development. By underlining the dissonant quality of the international response to the Brundtland report, this section sets up the ensuing discussion on coordination problems.

114 Brundtland to WCED commissioners, "Completion of the Commission's Mandate," October 12, 1987, Strong Collection – ESPP, Box 392, Folder 3734.

115 Lindner to Brundtland, "Follow-up on Commission's Report," August 24, 1987, Strong Collection – ESPP, Box 392, Folder 3740.

116 Kjellén 2008, 4.

Negotiating United Nations General Assembly Resolution 42/187

Before reaching the General Assembly, a draft resolution on the WCED report was negotiated at the June 1987 UNEP Governing Council session. These contentious negotiations hammered out the text that, with minor modifications, became resolution 42/187. At the Governing Council, two drafts formed the basis for negotiations: one from UNEP and another from the Nordic countries. The UNEP version bunched the WCED and the IIPC reports into one resolution, tying the two processes conceptually, legally, and politically. UNEP's draft gave the UNEP Governing Council the main responsibility for reviewing and reporting on long-term strategies to achieve sustainable development.[117] The Nordic version, on the other hand, rendered the WCED and IIPC reports into two separate resolutions. Consistent with the recommendations of *Our Common Future*, the Nordic draft gave the UN secretary-general the main responsibility for following-up on the institutionalization of sustainable development across the UN system, reporting directly to the General Assembly. This process, the Nordic countries suggested, ought to be facilitated by the inclusion of an item on the agenda of the 43rd General Assembly session, titled "Transition towards Sustainable Development," where the secretary-general would provide a preliminary report. To this end, the governing bodies of the organizations and programmes of the UN system should report to the secretary-general. To facilitate concrete action, the transition toward sustainable development would, again, be included on the agenda of the 44th General Assembly session.[118]

In addition to the negotiations held at the UNEP Governing Council, the resolution was debated within the UN's Economic and Social Council (ECOSOC) and at the General Assembly's Second Committee, as per General Assembly resolution 38/161.[119] The final

[117] United Nations Environment Programme 1987.

[118] "Nordic Draft General Assembly Resolution by the 14th Governing Council of UNEP on the Report of the World Commission on Environment and Development," May 25, 1987, 1144B/OSLO/bb/27.5.87-1, Strong Collection – ESPP, Box 392, Folder 3743.

[119] Some countries, especially Sweden, aimed to have the report reach the General Assembly directly. The fact that the special commission's report would first have to be passed by the UNEP Governing Council was another turf protecting

version of the resolution – passed in December 1987 as resolution 42/187 – recognized sustainable development as "a central guiding principle of the United Nations, Governments and private institutions, organizations and enterprises."[120] While the Nordic countries were able to secure separate resolutions for the WCED and IIPC reports, the follow-up provisions differed significantly from those outlined in their initial draft.[121] Efforts to place the secretary-general and General Assembly at the center of follow-up were almost completely rebuffed. Status reports from UN organizations and programs would be sent to the General Assembly via ECOSOC. The secretary-general's status report on implementation across the UN system would also reach the General Assembly via ECOSOC. These steps placed ECOSOC at the center of the process, thus minimizing the secretary-general's role in follow-up. Instead of an agenda *item* on "Transition towards Sustainable Development" at the 43rd and 44th General Assembly sessions, resolution 42/187 called for a *sub-item* titled "A long-term strategy for sustainable and environmentally sound development" at the 43rd General Assembly.[122] By treating sustainable development as a *sub-item*, rather than a full item, follow-up mechanisms were made less prominent and ensured that less time would be devoted to the topic. The use of a sub-item, furthermore, implied less urgency than the Nordic version.[123]

measure of resolution 38/161. Lars Björkbom Interview with the author, July 24, 2012.

120 United Nations General Assembly 1987.

121 Reflecting on the General Assembly's treatment in December 1987, the WCED secretary-general wrote to Commissioner Maurice Strong: "Personally, I hope we can view [UN General Assembly resolution 42/187] as no more than an initial, and perhaps necessarily very general, response to the Commission's recommendations … While I find it difficult not to be disappointed with the General Assembly's resolution, even as a first, general response, it does contain a few hooks on which to hang efforts for further action within governments, within regional bodies and with the specialized agencies of the UN. It could and should also lead to further consideration of the Commission's report and recommendations at the next sessions of the General Assembly." Jim MacNeill to Maurice Strong, December 28, 1987, Strong Collection – ESPP, Box 392, Folder 3744.

122 United Nations General Assembly 1987.

123 This analysis is somewhat at variance with the conclusions drawn by Iris Borowy (2014: 161) concerning the resolution. While she notes "real momentum" for "profound policy changes," I maintain that the resolution lacked specificity on follow-up.

The Puzzling Response to *Our Common Future*

When states met to address the recommendations of WCED, they faced mounting environmental problems. Many of these problems had been widely acknowledged through resolutions and conventions that aimed to halt environmental deterioration. Changes in key parameters rendered existing institutional arrangements unsuitable for addressing the environmental problems facing the world by the mid-to-late 1980s. UNEP lacked the resources or the authority to accomplish its coordinative mandate and was, by design, focused on the effects, rather than the causes, of environmental decline. UNEP's limited capacity to confront the changed international landscape was laid bare in the Brundtland report and had been acknowledged by the Intergovernmental Inter-sessional Preparatory Committee.[124] The UN's environment programme was, moreover, becoming increasingly marginalized by states. Influential governments – North and South – consistently sent low-level representation to Governing Council sessions and curtailed their political and financial investments in the programme.

Although the statements of key countries at the fall 1987 UN General Assembly session were decidedly cautious, their words nevertheless evidenced high receptivity to the type of institutional bargain proposed by the commission. Thirty-eight governments directly addressed the commission's findings in their plenary session speeches, welcoming the report and emphasizing its importance. Significantly, it found strong backing from influential developing world actors, such as Prime Ministers Robert Mugabe of Zimbabwe and Rajiv Gandhi of India.[125] Though he cautioned against environmental conditionality, Mugabe – who was then chair of the Non-Aligned Movement and a highly influential voice in the developing world – embraced the commission's report and backed its conclusions on the links between environment and development.[126] Welcoming the Brundtland Commission's condemnation of green conditionality and referencing the environmental disasters in Bhopal and Chernobyl, Gandhi was even more forthcoming. The prime minister pledged India's

[124] Engfeldt 2009, 114. Intergovernmental Inter-Sessional Preparatory Committee 1986.

[125] Lewis 1987, A5. [126] United Nations Press Release 1987.

"unstinting support" to realizing sustainable development.[127] The position of Brazil had been positive toward the commission since the WCED meeting in São Paulo in 1985 and Colombia pushed for the report to be "converted into an action plan towards a sustainable future."[128]

Statements from leading developed countries also indicated receptivity to a consideration of institutional change. While it would be too much to characterize the Reagan administration's position as one that was favorable to *Our Common Future*, its statements on the report were far less venomous than they had been toward other UN environmental initiatives. The presence on the commission of the influential Bill Ruckelshaus, Reagan's former Environmental Protection Agency administrator, likely contributed to this.[129] Though the United States was "not enamored" with some elements of the report, A. Alan Hill, the chair of the US Council on Environmental Quality, noted that "there are themes in that report that we are very supportive of."[130] Brundtland's elevation to the premiership of Norway, making her the leader of a close NATO ally, also contributed to a better reception to the report in Washington.[131] WCED's backing of growth-oriented economic policies was one such theme. Though it opposed the creation of new institutional machinery, the United States called for the implementation of sustainable development in the UN system.[132]

The British position, which was initially hostile to the commission, had become increasingly supportive. The British government hosted and paid for the launch of *Our Common Future* in London.[133] In an interview with the author, Sir Crispin Tickell, then the United Kingdom's UN ambassador, noted that the British position was more

[127] WCED Secretariat Memorandum, "What They Said about 'Our Common Future' – 'The Brundtland Report,'" Strong Collection – ESPP, Box 392, Folder 3734; United Nations Press Release 1987.

[128] MacNeill to Brundtland, "Meeting with the Intergovernmental Inter-sessional Preparatory Committee," March 12, 1986, Strong Collection – ESPP, Box 318, Folder 3015.

[129] This was aided by a softened attitude toward environmental issues in Washington. For an analysis of the evolution of US policy, see: Hopgood 1998, 125.

[130] Shabecoff 1987a, B12.

[131] Gro Harlem Brundtland Interview with the author, August 17, 2021.

[132] United Nations Press Release 1987.

[133] Dr. Jim MacNeill Interview with the author, July 2, 2013.

favorable than people assumed at the time.[134] At the special General Assembly session on environment and development, Tickell conveyed British backing for the integration of environmental considerations into development planning and called for a strengthening of UNEP's monitoring capabilities.[135] In 1988, the UK Department of Environment released a pamphlet detailing the government's perspective on the report.[136] Japan, which had been a principal financial and political sponsor of the commission, advocated the integration of environmental impulses with economic policies and called for the incorporation of sustainable development within multilateral financial institutions.[137] Canada pushed for the implementation of sustainable development across the UN system and offered to host an international conference on environment and sustainable development in 1992.[138] And the Nordic countries were, led by Brundtland herself, in the political lead on the institutionalization of sustainable development.

The statements of the Eastern bloc countries were among the most encouraging. The Soviet Deputy Minister for Foreign Affairs, V. F. Petrovsky, backed the commission's proposal for the negotiation of a global convention on environment and sustainable development that would ensure legal coherence on environmental questions. He noted, "the expanding of glasnost in these matters should become a part of the global efforts to defuse the environmental bomb." The Czech representative advocated a global action strategy for tackling existing and foreseeable environmental problems. Hungary highlighted the importance of the commission's proposals for strengthening UNEP.[139]

In addition to voicing their support for the implementation of sustainable development, some states took action in the ensuing years to embed elements of the Brundtland report in other, smaller multilateral fora where coordination problems were less severe.[140] The response to

[134] Sir Crispin Tickell Interview with the author, August 10, 2012.
[135] United Nations Press Release 1987.
[136] United Kingdom Department of Environment 1988.
[137] United Nations Press Release 1987.
[138] United Nations Press Release 1987. The Canadian Environmental Advisory Council also released a response to the commission report: Canadian Environmental Advisory Council 1987.
[139] United Nations Press Release 1987.
[140] The severity of coordination problems is, on average, less in institutions with smaller, more homogeneous memberships. The OECD, for instance, had twenty-four members in 1988 and the ECE had thirty-three.

Our Common Future within the OECD was rapid, for instance. In November 1988, a special high-level meeting was called to discuss the implications of *Our Common Future* for the organization. Participants at the meeting endorsed sustainability as a principle for the OECD's programs and argued that it should lead the way internationally by "elaborating the concept in economic terms." The OECD ministerial council treated sustainable development issues at their 1989 meeting, arguing that the concept should provide guidance on the organization's strategy and work program. Following the 1989 council meeting, the OCED's environment committee endorsed five themes for a medium-term program, including "the economics of sustainable development," "society, technology and environment," and "environmental aspects of development."[141] Similarly, the Economic Commission for Europe – which served chiefly as an East-West forum for cooperation – endorsed sustainable development goals at their ministerial meeting in Bergen, Norway, in May 1990.

Thus, there was a dissonant quality to the international response to the Brundtland report at the UN General Assembly. Governments widely embraced the report and recognized sustainable development as a central guiding principle in the UN system. This step was a departure from existing modes of cooperation. Governments agreed with most of the commission's diagnoses and took steps in other fora to realize sustainable development. However, their follow-up on the commission's recommendations for the United Nations system was cautious. While there was immediate recognition among leading states of the comprehensiveness of the sustainable development vision, for example, states effectively reserved their positions on institutional change by resorting to a "predictable competition about issue definition."[142] Instead of agreeing on concrete and decisive follow-up mechanisms, as proposed in the draft Nordic General Assembly resolution, states equivocated. Posturing, rather than purposeful negotiations, defined governmental responses to WCED in 1987.

The Brundtland Report and Coordination Problems

By the mid-to-late 1980s, change in environmental conditions had rendered existing institutional arrangements suboptimal and created

[141] Long 2000, 18, 22, 70–72. [142] Borowy 2014, 160.

incentives to realize institutional change. The report of the World Commission on Environment and Development provided a prominent diagnosis and a set of recommendations to bring institutional structures into alignment with the changed problem-setting. Although states were receptive to the commission's alternative agenda and took actions in some other intergovernmental bodies to institutionalize elements of the sustainable development perspective, their approach to reform in the UN system was decidedly cautious. All recognized that sustainable development was likely to become an important basis for UN cooperation, but states did not get down to the serious negotiations that would be required to operationalize the concept.

There was, as Marc Pallemaerts notes, an "artful vagueness" to the concept of sustainable development, which could only be clarified through negotiations.[143] Thus, beneath the surface of sympathetic General Assembly statements, governments anticipated more detailed bargaining in the future. The cautious, although apparently receptive, attitude of countries is explained by their desire to show an openness to talks, yet a reluctance to compromise their bargaining position in future negotiations by appearing too enthusiastic. While all could gain through institutional change, states faced important risks. If they were to act in a cooperative manner, revealing their underlying preferences, they could easily end up worse off than if they had continued to take a defensive posture in United Nations fora. This is a textbook illustration of the temporal variant of the negotiator's dilemma discussed in Chapter 1.

The developing countries had much to gain from institutionalizing sustainable development within the UN system. The Brundtland report stated bluntly that developed countries would have to take the lead part in financing sustainable development. Contributions through automatic financing – long a hobbyhorse of poorer countries – or through increased development assistance flows meant that the developing countries could benefit financially and environmentally.[144] The reliance of heavily indebted countries on development funds to prop up their wobbly economies in this period made the prospect of augmented multilateral and bilateral development assistance flows

[143] Pallemaerts 1994, 14. A similar view is taken in: Brenton 1994, 129; Kjellén 2008, 4.

[144] WCED 1987, 334–337.

highly attractive. The environmental and natural resource problems that could be ameliorated through cooperation were, moreover, genuine and urgent. The famine in Ethiopia and the Bhopal disaster illustrated the terrible results that environmental neglect and flimsy regulations could produce. If, furthermore, sustainable development made the United Nations a more prominent focus for cooperation, the international influence of well-represented developing countries would be enhanced.

At the same time, developing countries had a great deal to lose if sustainable development was institutionalized in a manner that was disadvantageous to their interests. While the Brundtland report called for heightened assistance levels, it emphasized the potential environmental benefits of doing so. The specter of "green conditionality," a deep concern at the 1972 Stockholm conference, reemerged in 1987. At the General Assembly, Nigeria and Mexico took the lead in warning about the potential dangers of embracing sustainable development as a basis for international cooperation. The Nigerian representative noted that the international community "should guard against translating concern for environmental protection into a new form of conditionality for development assistance by the multilateral development and financial institutions so that the cost of development will not be made unnecessarily more burdensome for the already overburdened developing countries."[145] Mexico strongly backed the Nigerian position.[146] Tied to green conditionality was the sensitive topic of sovereignty over resources and development planning. Any global action program, leading Latin American governments maintained, would have to take appropriate notice of the sovereign competence of governments to exploit their natural resources.[147]

The developed countries also faced risk. Public concern in the global North about environmental deterioration was widespread in 1987.[148] The urgent push to establish a basis for global cooperation to limit the emission of ozone destroying chlorofluorocarbons in the 1980s

145 "Extracts from Country Statements Made in the General Debate of the 42nd Session of the GA," Strong Collection – ESPP, Box 391, Folder 3733.

146 "Report on ECOSOC 1987: Resumé of Statements Made in the First (Economic) Committee of ECOSOC on Agenda Item 10 (International Cooperation in the Field of Environment), 1–2 July 1987," Strong Collection – ESPP, Box 392, Folder 3740.

147 Engfeldt 2009, 116.

148 Brenton 1994, 57.

demonstrated the necessity of engaging developing countries in global environmental issues.[149] The need for effective UN environmental institutions was unlikely to disappear with many pressing issues on the agenda, such as the loss of biodiversity, tropical deforestation, and climate change. At the same time, industrialized countries risked being obligated to assume an onerous financial burden in the pursuit of sustainable development, a still vague concept that could be operationalized as a developing world license to pollute and exploit their natural resources.[150] Ensuring that global responsibility for arresting planetary degradation was shared on a broad international basis was crucial to the efficacy of cooperative arrangements. As in 1972, protecting trade and investment interests was vital to safeguarding economic vibrancy. The rise of pollution havens would be economically and environmentally detrimental to the developed countries. Increased technological transfer to the developing countries could infringe on property rights.[151]

Embracing sustainable development was also risky for environment departments within national governments. The WCED report argued that environmental ministries would have to be strengthened significantly if sustainable development was to be realized nationally and internationally. This should involve a significant boost in the authority and resources devoted to environmental protection bodies, the report contended.[152] While sustainable development *could* aid environment ministries if governments directed more resources to managing environmental problems, there were uncertainties. *Our Common Future* argued that central agencies, such as finance departments, should play the lead part in ensuring sustainable development.[153] Environmental agencies, already underfunded, risked being further marginalized by the sustainable development agenda if they did not protect their bureaucratic flank. In an interview, Jim MacNeill recalled:

149 Brenton 1994, 142. See also: Speth and Haas 2006, 2–23; Tolba 1998, 62–74.

150 A United Kingdom Department of Environment report noted that automatic funding mechanisms "are not cost-free for countries and taxes on trade inhibit economic growth, which generates the resources on which both environmental protection and development rely." See: United Kingdom. Department of Environment 1988, 60–61.

151 United Nations Press Release 1987.

152 WCED 1987, 319–320.

153 WCED 1987, 314.

I was convinced of the need for the alternative agenda. I believed that the transition needed to come from heads of state and heads of government, and involve the powerful agencies driving environmental destruction ... Environmental agencies had assumed that they'd come out with a strengthened and new mandate. I couldn't let that happen. They had no responsibility for the policies in question ... after the manuscript [of chapter 12, on institutional change] got around, I got phone calls from deputy ministers of environment from a number of countries accusing me of betraying them.[154]

Bureaucratic politics within government administrations also contributed to the slowing of the international embrace of sustainable development. Environment departments were well represented at the UNEP Governing Council and influenced government negotiating positions at the UN General Assembly. As a result, many environment departments responded to the Brundtland recommendations with caution.

The United Nations Environment Programme faced similar risks as national environment departments. *Our Common Future* argued that UNEP's coordinative and catalytic mandate should be strengthened. UNEP's Environment Fund should be augmented and refocused to maximize its impact, argued the commission.[155] UNEP would be an indispensable part of the sustainable development enterprise within the United Nations system. If the WCED recommendations were followed, the environmental pie would expand significantly, benefiting the UN programme. Yet, like environment ministries, if countries did not act to boost their financial and political commitment to environmental institutions, UNEP could be left significantly worse off through embracing WCED's conception of sustainable development. The response, almost inevitably, recalled Brundland, produced "counterefforts by many who [tried] to protect the [institutional] present."[156]

All sides therefore faced a crucial risk: If they advocated sustainable development and other relevant actors did not, they could be left significantly worse off than had they bargained defensively. If developing countries embraced sustainable development policies by negotiating cooperatively, they risked legitimizing green conditionality. If developed countries negotiated cooperatively, they risked validating

[154] Dr. Jim MacNeill Interview with the author, July 2, 2013.
[155] WCED 1987, 319–323.
[156] Gro Harlem Brundtland Interview with the author, August 17, 2021.

patterns of development that could hasten and exacerbate global environmental decline. While national environment ministries and UNEP could benefit from the sustainable development agenda, they could also be marginalized bureaucratically without complementary action by decision-making and resource-endowing bodies. Rather than embrace sustainable development, therefore, cautious and defensive approaches prevailed.

Coordination problems thus hampered the move to institutionalize sustainable development within the United Nations system in 1987. Though existing institutional arrangements were suboptimal in the changed institutional environment, the risk-dominant character of the existing equilibrium worked to increase the resilience of the status quo. This inability to achieve broad-based coordination even when change is needed and urgent, recalls Brundtland, is "the reality of the UN."[157] At the root of this problem were issues of timing. The question was as much about *when* to negotiate cooperatively as it was about *whether* to do so. If one set of actors were, for instance, to reveal their underlying preferences to the others in the hopes of realizing agreement, they could be exploited. The key was for actors to move synchronously.

Temporal coordination was, however, problematic because, as noted above, there was considerable ambiguity about the timeline for follow-up on WCED's report. Though the arrival of *Our Common Future* was unique and highly conspicuous, no clear, unambiguous time frame emerged concerning institutional bargaining to implement its recommendations. As late as October 1987, Gro Brundtland believed that an immediate follow-up mechanism could be approved at the General Assembly's 42nd session.[158] Others anticipated a longer-term process, perhaps culminating in a still notional international conference in 1992. In an interview, Brundtland highlighted the challenges of ensuring consistent follow-up among leaders in realizing institutional change "as new issues arise and take up leader's

[157] Gro Harlem Bruntland Interview with the author, August 17, 2021.

[158] In an internal memorandum on October 12, Brundtland reported that she "had reason to believe" that "the General Assembly will decide on some follow-up process which will involve governments and the UN itself as well as its specialized agencies." See: Brundtland to WCED commissioners, "Completion of the Commission's Mandate," October 12, 1987, Strong Collection – ESPP, Box 392, Folder 3734.

daily agendas."[159] The result was a divergence, rather than a convergence, of expectations. Because the best response was for actors to behave in the manner that they expected others to behave, a cautious response was demanded by the situation. Thus, despite the availability of a payoff-dominant equilibrium, the status quo was maintained.

In the absence of a large international conference, as with the other cases considered in this book, should reformers have been surprised by the lack of serious bargaining in 1987? The UN General Assembly and its associated committees can, and regularly do, provide a forum for institutional bargaining that could have enabled the implementation of WCED recommendations. While a special conference could also be convened, governments were well-positioned to negotiate in fall 1987 on the set of recommendations provided by WCED. The recommendations had been circulated well in advance of the General Assembly session and were considered by members of the commission to be highly urgent. As a theoretical matter, in the presence of available incentives to update institutional mechanisms, we should expect governments to move expeditiously to take action to seize those incentives.

Conclusion

Judged from the perspective of UN General Assembly resolution 42/187, efforts to precipitate change in UN environmental governance on the basis of the Brundtland report did not succeed. Despite their openness to the institutional bargain sketched in *Our Common Future*, states did not engage seriously in the task of reforming UN environmental institutions in 1987. The glaring institutional, environmental, and economic problems that were stressed by the commission – and acknowledged by states – were left unaddressed. States' tentative response to the commission's report at the 1987 General Assembly session is made more perplexing by the fact that they institutionalized sustainable development in other international fora and would accept many of the commission's recommendations less than five years later.

The international response to *Our Common Future* is crucial for our understanding of the role of temporal coordination in institutional change. Though the report had a high-profile and a unique standing when it reached the United Nations General Assembly, there was

[159] Gro Harlem Brundtland Interview with the author, August 17, 2021.

ambiguity in the timeline for follow-up actions associated with it. This ambiguity contributed to a temporal divergence of expectations concerning negotiations over the operationalization of sustainable development across the UN system. Divergent expectations hindered greatly the realization of institutional change because of temporal coordination problems. While all actors stood to gain from achieving sustainable development, they also faced important risks. If one side behaved in a cooperative manner and others did not, the cooperative player risked winding up significantly disadvantaged in future negotiations and, as a result, bargained defensively. Posturing, rather than open negotiations, resulted.

As we will see in Chapter 6, the 1992 Rio Earth Summit embraced many of the recommendations of the World Commission on Environment and Development.[160] Since the international problem structure was largely constant between the two case studies, the Brundtland and Rio cases are a good and close comparison. The principal difference is that the Rio conference constituted a Temporal Focal Point in UN environmental affairs, facilitating a temporal convergence of expectations. The divergence in outcomes argues for the vital role that temporal signposts play in solving coordination problems.

[160] Kjellén 2008, 4.

6 *The Rio Conference and Institutional Change*

The late 1980s and early 1990s were a high-water mark for world environmental concern. Although follow-up on the recommendations of the World Commission on Environment and Development (WCED) was hobbled by temporal coordination problems in the UN context, the General Assembly negotiations over a global conference to mark the twentieth anniversary of the 1972 Stockholm conference provided an avenue for the realization of change. United Nations General Assembly resolution 44/228, which launched the United Nations Conference on Environment and Development (UNCED), set in motion a process that would lead to the adoption of key WCED recommendations and significantly alter existing modes of global environmental cooperation.

UNCED, which was held in Rio de Janeiro, Brazil, from June 3 to 14, 1992, was an event of unprecedented scale and scope. The Rio conference was attended by more than 178 countries, virtually all of the agencies and programmes of the UN system, and by representatives of 35 other intergovernmental organizations. As many as 108 heads of state or government attended the summit, making it the largest gathering ever of world leaders. A total of 7,000–8,000 delegates represented their countries, and 9,000 journalists reported on the event. Between 1,400 and 1,450 nongovernmental organizations were accredited at the UN conference and the parallel nongovernmental "Global Forum" was attended by 20,000 environmentalists. All told, 40,000–45,000 people in various capacities descended upon the South American country for the conference.[1] The Brazilian meeting resulted in a conference declaration, an action plan, a framework convention on climate change, a convention on biological diversity, and a set of forest principles. Among the recommendations contained in "Agenda 21," the conference action plan, were those that recommended the

[1] Engfeldt 2009, 187–188; Shabecoff 1996, 152–153.

creation of a Commission on Sustainable Development and formalized a role for the World Bank-operated Global Environment Facility (GEF) in post-UNCED follow-up. The provisions in the Rio declaration and action plan instituted a revised vision of cooperation within UN environmental affairs, integrating environmental goals with social and economic ends. The "Earth Summit" led to significant institutional change.

This chapter begins with a summary of major developments from 1987 to 1992, assessing some of the challenges and limitations facing UN environmental institutions in this period. I consider the state of the world environment, determining the extent to which shifts in global conditions provided incentives for change. I then provide a summary of the principal benefits and risks facing states in realizing change. Next, I demonstrate that the Rio conference constituted a Temporal Focal Point in the history of UN environmental politics, providing a renewed impetus for negotiations to bring institutional structures into closer alignment with underlying conditions.

Changing International Conditions and the Rio Conference

Shifts in the world economic situation through the 1980s had altered key parameters in the UN environmental sphere. The Third World debt situation reached crisis proportions following the threat of a Mexican government default on its external public debt in August 1982.[2] The pool of new private lending available to developing countries rapidly dried up.[3] Low commodity prices greatly diminished export earnings.[4] These factors, combined with high interest rates, made debt levels in many developing countries unmanageable. By 1988, the debt of African countries was nearly US$140 billion.[5] Between 1982 and 1992, Latin American countries transferred US$220 billion to the rest of the world, in large part because of onerous

[2] On Mexican macroeconomic policy and prospects at the time, see: Rojas-Suarez 1992.

[3] Private banks held more than half of the US$1.3 trillion Third World debt. For a detailed treatment, see: Culpepper 1991, 11.

[4] Between 1980 and 1984 there was a loss of about $55 billion in export earnings for developing countries. See: WCED 1987, 36.

[5] Stein and Nafziger 1991, 174.

debt-service payments.[6] During the 1980s, many Latin Americans experienced a 35 percent decline in real per capita income.[7]

Third World debt had profound environmental consequences because about 70 percent of developing world debt was owed by states that relied on primary commodities for the majority of their exports. In Sub-Saharan Africa, for instance, primary products accounted for over 90 percent of exports.[8] As a result, many developing countries were forced to exploit their natural resources unsustainably as a means of servicing debt. Some examples are illustrative. In many African countries, cash crop cultivation increasingly supplanted food crop production. In addition to worsening poverty, cash crops led to agricultural policies that targeted short-term exigencies rather than long-term priorities. Crop rotation was widely neglected and forests were cleared for agricultural use, contributing to desertification and deforestation.[9] This situation led United Nations Environment Programme (UNEP) to conclude at the time of UNCED that "the pace of environmental degradation seem[ed] certain to increase unless the debt crisis [was] resolved and greater equity introduced to the world's commodity markets."[10]

Institutional approaches that failed to integrate environmental, economic, and social imperatives were unable to meet changed international conditions. Efforts to protect forests or biological diversity, for instance, needed to take social and economic factors into account if they were to be effective. Forest preservation policies had to recognize that many countries relied on fuel wood for the bulk of their energy needs. Similarly, global schemes to protect biodiversity had to account for the reliance of many developing world societies on wildlife and forest products to meet their dietary needs.[11] Because socio-economic policies were at the root of global environmental deterioration, a focus only on the *effects* of pollution could not address the basic *sources* of environmental problems.

[6] Bulmer-Thomas 1992, 1. [7] Runnalls 1991, 2.
[8] Tolba and Biswas 1991, 12. [9] ENDA-Tiers Monde 1991, 4.
[10] Tolba et al. 1992, 816.
[11] In fact, at the time, 90 percent of the energy needs of Nepal, Tanzania, and Malawi came from firewood and dung. Many more societies relied on these energy sources for 80 percent of their energy needs. Many African countries, such as Botswana, Nigeria, and Ghana were heavily reliant on animal protein. See: Tolba 1998, 129.

Addressing global problems, such as ozone depletion or climate change, furthermore, required North-South cooperation. Widespread indebtedness and meager private capital flows meant that international financing was essential to support preventative action on global threats.[12] Without this support, most developing states lacked the financial, institutional, and technical wherewithal to participate effectively in international environmental agreements. This recognition was central to the creation of the US$240 million multilateral fund to promote chlorofluorocarbon alternatives in developing countries and to the decision to launch the Global Environment Facility in 1990.[13] Much more was required, however. On climate change, the developed countries could not solve the climate problem in the long run without the developing world. By 2050, projected warming would be 40 percent higher without developing country participation than with it.[14] China alone accounted for approximately 14 percent of global carbon dioxide emissions.[15] Solutions, both at the national and international levels, had deep social and economic implications, requiring the multisectoral approaches advocated by WCED.

Although UNEP had played an indispensable part in catalyzing international action to protect stratospheric ozone in the mid-1980s, this outcome was somewhat of a double-edged sword for the UN programme.[16] For developing countries, the ozone agreement cemented the image of UNEP as a technical body concerned principally with the problems of rich societies.[17] As Tommy Koh, chair of the UNCED preparatory committee, recalls, UNEP was considered "too

[12] Thacher 1991, 67.

[13] The Global Environment Facility was launched in 1990 as a joint project of the World Bank, UNEP, and the United Nations Development Programme. See: Brenton 1994, 142.

[14] Lashof and Tirpak 1990, 40–43.

[15] Bergensen and Sydnes 1992, 41.

[16] On the ozone accords, see: Haas 1992b, 187–224; Litfin 1995.

[17] It is interesting to note the impact of the ozone accords on UNEP's status. While the accords generated some suspicion in the developing world, many saw them – not incorrectly – as the UN programme's greatest achievement. In the second edition of Lynton Keith Caldwell's *International Environmental Policy*, he wrote: "As of 1990, my evaluation of UNEP differs significantly from that offered in the first edition of this book. At that time, the limitations and difficulties encountered by UNEP received greater weight than I give them today ... Two factors largely account for the progressive change: first, the energetic and persuasive leadership of UNEP's executive director, Mostafa Kamal Tolba; second, the coming-of-age of the international environmental movement." Caldwell 1990, 81–82.

pro-environment and insufficiently pro-development."[18] The proliferation of multilateral environmental agreements on specific issue areas – such as ozone and hazardous wastes – also increased the level of fragmentation in UN environmental cooperation. Governments, one observer recalls, came to the conclusion that they were not in full control of the vast new regime complex.[19] As the main funders of international environmental action, developed countries became concerned to rationalize the increasingly "untidy pattern" of institutional measures.[20] This uncoordinated and, to some extent, competitive setup could badly undermine global efforts to achieve sustainable development.[21]

UNEP continued to suffer from the organizational problems that had plagued it since the 1970s. It had great difficulty coordinating the environmental work of larger UN specialized agencies, such as the UN Food and Agriculture Organization (FAO) and the World Health Organization (WHO). UNEP was simply too small, too institutionally weak, and too poorly funded to guide the policies of those bodies, a major element of its mandate. Its distant location in the Kenyan capital added to this problem. When asked about UNEP's difficulties, one interviewee put it baldly: "UNEP is in the wrong place."[22] In addition to the challenges associated with coordinating from afar, many developing countries did not see Nairobi as a favorable location to conduct high-level international business. Instead, these countries preferred New York because they tended to have higher-caliber and better-resourced diplomatic representation there.[23] Funding problems also dogged UNEP. Taking inflation into account, UNEP's budget in 1990 was almost identical to its budget in 1972. To put UNEP's funding problems into context: In the first *fifteen years* of its operations, total contributions to UNEP were less than half of the *annual* budget of FAO, WHO, and the UN Educational, Scientific and

[18] Tommy Koh Interview with the author, August 21, 2017.
[19] Engfeldt 2009, 139. [20] Brenton 1994, 260.
[21] David Runnalls commented in a paper that: "Because of the weakness of the central UN secretariat and the conflicted goals of governments, the specialized agencies and their heads have come to behave like medieval baronies, feuding among themselves and with the centre. This situation is especially damaging to the prospects of sustainable development, requiring as it does the integration of both environmental and economic concerns." Runnalls 1991, 9.
[22] Sir Crispin Tickell Interview with the author, August 10, 2012.
[23] Brenton 1994, 185.

Cultural Organization.[24] The programme's voluntary funding structure simply was not working, limiting its ability to coordinate world environmental action.[25]

UNEP's limitations were not new. They had been widely acknowledged at the 1982 Session of a Special Character and were again highlighted in 1987 when the Brundtland report was released. Yet these organizational defects had been insufficient to precipitate institutional change. Hackneyed calls to address UNEP's many limitations failed to impress decision-makers. This was still the case in the lead-up to Rio. One observer wrote at the time that the current order was "incapable of coping with the new set of problems which are likely to dominate the UN agenda for the rest of the century."[26] Exogenous shifts had, moreover, created incentives for a closer integration of economic and environmental decision-making. UNEP was incapable of achieving this.

The changed international setting manifested itself in the attitude of countries toward preparations for UNCED. Detailed negotiations were held in 1988 and 1989 over the nature of conference preparatory processes. UNEP and its executive director, Dr. Mostafa Tolba, were cool to the conference idea from the beginning and sought to limit the feared damage to UNEP.[27] Tolba lobbied extensively to serve as conference secretary-general and to have the UNEP Governing Council function as the UNCED preparatory committee.[28] These measures would place UNEP at the center of the UNCED process and position it to assume major conference follow-up responsibilities. The new, broader global environmental problem-setting, however, made UNEP unsuitable for this role.[29] A much wider set of political considerations would need to be taken into account, encompassing economic and social factors. UNEP was not well-positioned to manage such a central and encompassing task.[30]

[24] Munro 1991, 18.
[25] For a detailed account of UNEP's funding challenges, see: Ivanova 2012.
[26] Runnalls 1991, 8.
[27] Tolba was irked to discover in autumn 1988 that UNEP had not been mentioned in an early draft General Assembly resolution to convene the conference; the resolution had been prepared by leading reform-minded countries. See: Engfeldt 2009, 119.
[28] Engfeldt 2009, 118, 124–125.
[29] Engfeldt 2009, 122.
[30] Sir Anthony Brenton Interview with the author, July 19, 2012.

The result of these negotiations was a delicate formula that established a preparatory committee of the General Assembly. Of the four formal preparatory committee meetings, one would be held in New York, another in Nairobi, and two were to be convened in Geneva.[31] While some effort was made to preserve UNEP's standing in General Assembly resolution 44/228, there was little doubt that the programme had been marginalized. Tolba was not, moreover, selected as conference secretary-general.[32] Instead, UN Secretary-General Javier Perez de Cuéllar chose Maurice Strong, who had served in the same capacity at the Stockholm conference.[33] Among other factors, Strong's selection may have been the result of a Brazil-Canada deal in which Canada withdrew its offer to host UNCED in return for Brazilian support for appointing Strong as conference secretary-general.[34] Tolba may also have been outmaneuvered bureaucratically. In the recollection of one observer, once appointed, Strong "wanted to be the leader and he was not prepared to share the limelight with Dr Mostafa Tolba."[35]

Significant changes in the international setting had altered the cooperative basis for United Nations environmental institutions. Economic crisis in the 1980s contributed to a worsening of world environmental conditions. Current institutional arrangements – including UNEP and a loose complex of multilateral environmental agreements – were ill-suited to the new context, creating incentives for a transformation of behavioral norms, rules, and procedures. The "react and retrofit" approach to planetary management failed to address the underlying causes of environmental decline.[36] The new situation

[31] United Nations General Assembly Resolution 44/228 1989.

[32] Hopgood 1998, 169–170.

[33] For Maurice Strong's account of the appointment, see: Strong 2001, 188–189. In addition to his institutional position, Tolba was seen as unsuitable because of his controlling management style. In an interview, Sir Martin Holdgate recalled. "UNEP was seen as Tolba's fiefdom. We needed to shake the cage …" (Sir Martin Holdgate Interview with the author, August 6, 2012). Similarly, Lars-Göran Engfeldt commented in an interview that Tolba was "too authoritarian." He continued: "Tolba's tactics were not that sophisticated and he alienated people. Tolba tried to get the upper hand and sidelined UNEP himself by refusing to cooperate with the Rio secretariat." (Lars-Göran Engfeldt Interview with the author, July 23, 2012).

[34] Corrêa do Lago 2009, 140.

[35] Tommy Koh Interview with the author, August 21, 2017.

[36] WCED 1987, 7.

demanded institutional measures that safeguarded the environment in a manner that took economic and social factors into account.

The Rio Negotiations

In the years preceding the 1992 Rio conference, exogenous shifts had altered key parameters in the UN environment sphere. Existing institutional measures were poorly suited for the changed problem-setting, bringing to light the need for a closer integration of environmental, economic, and social factors. This section elaborates on the interests of actors and on the coordination dilemmas that they faced in realizing accumulating incentives for change. Particular emphasis is placed on the risk-dominance of the prevailing equilibrium.

The Payoff-Dominant Equilibrium

After its release in 1987, WCED's report, *Our Common Future*, provided the most prominent conceptual framework for institutional reform.[37] While UN General Assembly resolution 42/187 recognized sustainable development as "a central guiding principle of the United Nations, Governments and private institutions, organizations and enterprises," follow-up mechanisms on the WCED recommendations in the UN context were unspecific.[38] At the root of states' cautious approach were temporal coordination problems. Governments, North and South, did not commit themselves to the sustainable development concept too quickly, lest they compromise their bargaining position in future negotiations. Ambiguity over negotiating timelines led to a temporal divergence of expectation and contributed to states' defensive posture in talks.

The cautious reception that the WCED report received at the UN General Assembly should not be interpreted as a rejection of the commission's recommendations or of sustainable development. As noted in Chapter 5, a review of the reception of leading actors to the Brundtland report indicates receptivity to the commission's institutional change proposals. All sides indicated a willingness to bargain and established their principal negotiating points. While UN talks were at a standstill, states moved to implement sustainable development

[37] WCED 1987. [38] United Nations General Assembly 1987.

policies in several other, smaller fora. At the Organisation for Economic Co-operation and Development (OECD), a developed world forum, member states instituted many elements of the sustainable development agenda.[39] Implementation was also advanced within some governments. Canada introduced a National Roundtable on the Environment and the Economy in 1988 and the United Kingdom appointed junior environment ministers to every government department.[40] At the September 1989 Tokyo Conference on the Global Environment and Human Response Toward Sustainable Development, the government of Japan committed to earmarking US$2.25 billion of development assistance for environmental purposes.[41] Nordic countries embraced the Brundtland Commission's message and several African governments began to implement sustainable development policies domestically.[42] Despite some progress within national governments and certain international fora, institutional bargaining – and the political and analytical investments associated therewith – within the wider UN system did not progress. Where coordination problems were less severe and transaction costs were lower, implementation proceeded.

Though they continued to stress the primacy of economic development ends, many developing countries saw important benefits in achieving sustainable development on a global basis. As noted above, the Third World debt crisis led to a dramatic worsening of environmental conditions in much of the developing world. In 1991, forty-one developing countries recognized environmental decline as a "common concern of humanity."[43] Environmental measures were, it was increasingly acknowledged, central to achieving lasting and human-centered development.[44] Environmental catastrophes of the 1980s, such as the Bhopal industrial accident and the Ethiopian famine, brought attention to the importance of environmental policies to developing countries.[45] There was a growing recognition that environmental stresses disproportionately affected the poorest and most vulnerable societies.[46] The

[39] Long 2000, 18, 22, 70–72. [40] Runnalls 1991, 7. [41] Thacher 1991, 66.
[42] Dr. Jim MacNeill Interview with the author, July 2, 2013. Bob Munro Interview with the author, July 3, 2013.
[43] This comes from the Beijing Declaration of forty-one Developing Countries, quoted in: Hurrell and Kingsbury 1992, 44.
[44] Agarwal 1990, 94–96. [45] Williams 1993, 19.
[46] ENDA-Tiers Monde 1991, 3.

increasing public concern was reflected in the growth of environmental nongovernmental activism in the global South. These organizations gained public notoriety and drew attention to the development/environment relationship.[47] The 1991 Tlatelolco Platform on Environment and Development acknowledged the right of environmental nongovernmental organizations to involve themselves in domestic and international issues.[48]

Developing countries faced important limitations in their efforts to ameliorate environmental problems. Economic crisis reduced greatly the capacity of governments to target environmental problems, causing them to rely heavily on international funding sources. On this front, existing institutional arrangements were clearly not up to the job.[49] Initiatives to tackle desertification, which continued to be an acute problem in much of the developing world, were hampered by funding limitations.[50] International efforts to address global problems, such as ozone depletion and climate change, relied chiefly on the actions of the developed countries. There were thus benefits to letting industrialized countries take the lead in addressing these perils, even if these environmental problems were also detrimental to world development prospects. In addition to potentially gaining additional financial resources to improve environmental conditions, developing countries coveted technological transfers that could enable them to reduce pollution and industrial waste. Access to more efficient technologies had the added bonus of making poor countries more competitive economically.[51]

Developed countries also sought to realize important benefits. The acceleration of environmental deterioration in the global South motivated industrialized countries in their pursuit of institutional change.

47 Marc Williams cites the Chipko movement in India and the Amazon rubber tappers as groups that gained national and international media attention for their activities. See: Williams 1993, 16.

48 "Letter Dated 20 March from the Permanent Representative of Mexico to the United Nations Office at Geneva Addressed to the Secretary-General of the UNCED (Tlatelolco Platform)" 1991, 446–450.

49 Thacher 1991, 62.

50 A Dutch report in 1990 showed that about 15 percent of the earth's surface had been degraded by human activities, such as overgrazing, deforestation, agricultural overexploitation, or bioindustrial activities. See: International Soil Reference and Information Centre 1990.

51 Hurrell and Kingsbury 1992, 41.

Rapid deforestation, spurred in large part by unsustainable development practices, was a growing problem in the 1980s and early 1990s. Attention was particularly focused on tropical forests, which were threatened by the spread of cash crop cultivation and the export of forest products. The United States took the lead in advocating an international forest convention that would be ready for signature in Rio.[52] Concern for world forests contributed to the popularity of "debt for nature swaps" in this period.[53] An associated, yet distinct, question at the time was the decline of world biological diversity. In 1992, UNEP reported that a quarter of the world's biodiversity was at risk of extinction in the next twenty to thirty years.[54] The loss of biodiversity motivated the Swedish government to take the lead – with Kenyan backing – in negotiating a UN Convention on Biological Diversity. These and other problems were of wide international significance and, because they largely centered in the global South, they demanded international cooperation.

There was also deep concern among developed states about threats to the global commons. More than 70 percent of the earth's atmosphere and oceans existed outside of national jurisdictions.[55] Many planetary problems, therefore, had to be tackled on a broad international basis if they were to be addressed effectively. Of these issues, climate change was the most distressing. At the time, atmospheric concentrations of carbon dioxide and methane were increasing by 0.5 and 0.9 percent per year respectively.[56] If a firm international basis for action was to be established, the participation of industrialized countries *as well as* large developing countries, such as China, Brazil, and India, was a must. China, for example, contributed more than 14 percent of global emissions of carbon dioxide, and had plans to greatly increase its energy consumption.[57] Finding a means of replacing China's 750,000 inefficient industrial boilers was, for example, absolutely necessary if the world was to slow global warming. The problem, as David Runnalls observed, is that, without

52 The United States enjoyed little success in this endeavor and had to settle for a set of forest principles that could form a basis for future negotiations. See: Hopgood 1998, 135.

53 For a discussion, see: Susskind 1994, 84–86.

54 Tolba 1992, 77.

55 Strong 1972, 115.

56 Tolba et al. 1992, 25.

57 Bergensen and Sydnes 1992, 41.

international financing, China had few incentives to bear these costs unilaterally.[58]

Although the price of achieving sustainable development on a global basis was steep, there were clear draws to altering current institutions as a means of slowing planetary decline. If a sound international basis for environmental cooperation was established, the South would be able to better manage growing environmental problems. An institutional transformation might have allowed them to attract greater resource flows and, potentially, negotiate the transfer of more efficient technologies. In the North, concern about the rapid deterioration of world environmental conditions caused these countries to seek international cooperation to arrest pressing problems, such as climate change or the disappearance of tropical forests and biological diversity. Institutional alternatives were *payoff-dominant.*

The Current Risk-Dominant Equilibrium

Despite incentives, states were slow to engage with the institutional recommendations of the Brundtland Commission. This caution was reflected in the fact that leading developed and developing countries resisted Norwegian efforts to use the words "sustainable development" in the UN General Assembly resolution stating the Rio conference's title. Instead, they opted for a United Nations Conference on *Environment and Development.*[59] They were reluctant to embrace the sustainable development agenda because, unless all relevant actors did so concurrently, they risked weakening their bargaining position in future talks. In this, states faced severe temporal coordination problems. Thus, while institutional change was payoff-dominant, great uncertainty about the actions of other players in the international system contributed to the persistence of the existing, *risk-dominant* equilibrium. Retaining the environment/development juxtaposition, rather than endorsing the fuller integration of these two issue areas within the sustainable development formula, was a safer bet for all concerned.

Developing countries faced numerous risks if they were to engage cooperatively when developed countries did not. If sustainable

[58] Runnalls 1991, 3.

[59] For an account of these negotiations, see: Engfeldt 2009, 118–121.

development were institutionalized in a manner that was insensitive to the needs of the developing world, change could easily be detrimental to their economic prospects. Through the 1980s, standards of living in Latin America, Africa, and much of Asia had deteriorated sharply. In Africa, per capita food production was in decline, infant mortality rates were on the rise, life expectancy had dropped, and the AIDS crisis continued. Economic crisis in Latin America brought discontent and threatened many of the newly established democratic governments in the region.[60] There were, therefore, good reasons for developing countries to avoid any measures that could hinder their development. Environmental programmes would, in many cases, be costly and it was in the interest of these countries to avoid expensive international commitments. Just as importantly, the capricious green impulses of developed countries could, they worried, easily divert financial and political resources from the international development agenda.

Developing countries were also apprehensive that sustainable development was a Trojan Horse for aid conditionality that might enable donor countries to interfere in their domestic policies. Conditionality was an increasingly popular tool in this period, much to the chagrin of aid beneficiary countries. If institutionalized in a manner that did not safeguard developing country interests, sustainability could be imposed as a condition of development assistance.[61] This fear seemed especially real in the early 1990s, as the World Bank had been placing increased emphasis on the sustainable development agenda. Environmental conditions were, moreover, the subject of a series of proposals in the US Senate in 1989 to link World Bank funding decisions to their anticipated environmental effects.[62] This talk was particularly objectionable in Brazil, which had significant economic need and unique environmental features, such as the Amazon rainforest.[63] The specter of green conditionality added a degree of uncertainty to development planning.

Developing countries also feared the possible use of environmental justifications for discriminatory trade practices, such as the imposition of nontariff barriers to trade. Many feared that access to vital markets could be restricted when the environmental standards of exporting

[60] Runnalls 1991, 2.

[61] Preparatory Committee for the United Nations Conference on Environment and Development 1991.

[62] Hurrell 1992, 406–407. See also: Binswanger 1987.

[63] Engfeldt 2009, 120.

countries were less stringent than those of importing states. These anxieties seemed justified. In the years before UNCED, a number of European municipalities banned the import of tropical timber in response to pressure from environmental groups.[64] The relationship between environmental safeguards and norms of open trading, such as those embodied in the General Agreement on Tariffs and Trade, remained ambiguous in the early 1990s.[65]

It was not just the developing world that approached the sustainable development agenda with caution. If sustainable development was to be institutionalized in a manner that gave too much weight to the development agenda, the Rio conference could legitimize ecologically destructive patterns of economic growth. If environmental aims were subverted in institutional change, the environmental consequences could be severe. It was these concerns that led some developed countries to advocate aid conditionality in their promotion of sustainable development.[66]

Developed countries risked being obliged to assume an onerous financial burden in support of international efforts that might be of limited relevance to safeguarding the world environment. Lead industrialized countries expressed concern that, if their environmental priorities were sidelined at Rio, the conference could turn into little more than a fundraising event of significantly less interest to them.[67] The early 1990s were, moreover, a time of economic recession in North America and the United Kingdom, leaving few resources for expensive international and environmental initiatives.[68] Similarly, Germany was absorbing the astronomical costs of reunification and most western governments were dealing with the economic ramifications of promoting the peaceful dismemberment of the Soviet Union and its former satellite countries.[69] Although France, Japan, and Sweden were contemplating sizable financial contributions to realize global sustainable development, this potentiality could not form a sufficient basis for such dramatic governance change. Maurice Strong, the conference secretary-general, is reported to have quipped: "Never have the rich felt so poor."[70] It was risky for wealthy countries to invest substantial

[64] Runnalls 1991, 3. [65] Hurrell and Kingsbury 1992, 46.
[66] Hurrell and Kingsbury 1992, 45–46. See also: Spector 1994, 87–103.
[67] Engfeldt 2009, 120. [68] Brenton 1994, 220. [69] Bruce 1991, 14–15.
[70] Engfeldt 2009, 143.

political and financial resources into an institutional project that might not address the global problems that they set out to solve.

Economic and sovereignty concerns were also present in leading developed states. In an election year, the administration of US President George H. W. Bush faced enormous political pressure to tighten budgets and resist perceived international profligacy.[71] The White House Council on Competitiveness – chaired by Vice President Dan Quayle – became involved in fighting UNCED initiatives that, in its view, compromised the United States' economic competitiveness.[72] Fear of developing world pollution havens, technological transfers that violated intellectual property rights, and costly impediments to the exploitation of biotechnological resources were major issues during the UNCED preparatory process. Consequently, major developed countries worked assiduously to minimize the position of international trade issues on the Rio agenda.[73] Although they were less vocal, many developed countries also refused to surrender authority over environmental rulemaking or rule implementation. The United States, in particular, was at least as reluctant to cede its sovereign rights to international authorities as many developing countries were.[74]

These risks caused developed and developing countries to approach institutional change with considerable caution during the UNCED preparatory process. As in 1987, coordination failure would result in a failed attempt at reform, wasted resources, and, for cooperative players, a weakened bargaining position in future negotiations. The timing issue was the key stumbling block from 1987 to 1992. Although important details had to be fleshed out by states, the basic institutional alternative had been clearly established since 1987. As we shall see, the character of institutional change realized in Rio was well in line with that prescribed by *Our Common Future*. Far more problematic was ensuring that states timed their embrace of the WCED formula so as to avoid compromising their bargaining position. No side wanted to

[71] The lead US negotiator reflected: "I think Rio would have been far more successful if it hasn't happened in a U.S. election year and, obviously, if it hasn't happened in a world recession period."

[72] The Competitiveness Council was especially active on the Biodiversity Convention negotiations. See: Hopgood 1998, 142, 171–174.

[73] Department of External Affairs and International Trade and the Department of Environment 1991, 4.

[74] Hurrell and Kingsbury 1992, 45.

commit too early. It was best for each to cooperate only *when* they expected others to do so. It was only when this uncertainty was reduced that states moved decisively to realize institutional change.

Since the distributive implications of talks were highly important, neither side wanted to signal weakness by appearing too eager to negotiate.[75] Delay tactics were a costly signal of strength and resolve. Appearing keen to negotiate signaled the reverse, harming the bargaining position of the more enthusiastic side in future discussions. By foregoing the benefits of Pareto-improving institutional change, actors signaled that they must be given a good deal in order to reach a settlement. Although both sides broadly supported the sustainable development formula, agreement on details would have to be reached on *their* terms. This contributed to the resilience of the risk-dominant, status quo equilibrium.

The Rio Conference and Institutional Change

The World Commission on Environment and Development advanced several far-reaching proposals for institutional change. The commission placed great emphasis on the need to integrate fully environmental endeavors with social and economic ends. The sectoral treatment of environment – separated from the decision-making processes that led to environmental problems – was the chief obstacle to realizing sustainable development. Within the UN system and elsewhere, environment was institutionalized in a manner that focused on the *effects*, rather than the underlying *causes*, of environmental decline.

The institutional negotiations that ultimately led to the creation of the UN Commission on Sustainable Development (CSD) canvassed a broad array of institutional options. At a spring 1991 working group meeting in New York, states discussed various means of altering institutional arrangements to account for the changed problem-setting. Among other options, they considered modifying UNEP's mandate, developing a closer association between UNEP and the United Nations Development Programme, transforming the rudderless UN Trusteeship Council into an environmental body, and resurrecting the defunct

[75] James D. Fearon has extensively discussed the role of signaling problems in war of attrition games. See: Fearon 1994, 1998, 284.

Environmental Coordination Board.[76] Despite entertaining a wide range of options, by spring 1992, the menu of institutional alternatives had narrowed significantly. Negotiations centered on a set of institutional proposals remarkably similar to those recommended by the Brundtland Commission.[77]

The CSD concept reached the conference preparatory committee as a "non-paper" from Malaysian Ambassador Ismail Razali, who was vice chair of the UNCED Working Group on legal and institutional questions.[78] Razali's draft argued that the UN General Assembly should be at the center of UNCED follow-up and envisioned a central coordinative role for the Administrative Committee on Coordination (ACC). Because the UN secretary-general headed the ACC, this proposal placed the United Nations secretariat at the heart of UNCED follow-up. The element of Razali's paper that attracted the most attention was the establishment of a Sustainable Development Commission, which would monitor progress on the implementation of Agenda 21.[79] By the end of the fourth preparatory committee meeting, states debated two basic visions of the commission concept. The first was a high-level commission that would report directly to the General Assembly on policy questions and to the UN's Economic and Social Council (ECOSOC) on coordination issues. The second version envisioned a similar body, except that it would function completely under the auspices of ECOSOC.[80]

The other major area of institutional debate during the UNCED process concerned funding arrangements for the implementation of Agenda 21. Two main options were presented. Developed countries sought to make the Global Environment Facility, established as a pilot

[76] Preparatory Committee for the United Nations Conference on Environment and Development 1991.

[77] The CSD proposal arose in the Rio context through a pair of informal sessions. It was first raised in the UNCED setting at an informal secretariat-organized seminar on institutional questions, hosted by the Aspen Institute in July 1991. At Aspen, the concept was initially raised by British diplomat John Martin, and was backed by the session's chairperson, Sir Crispin Tickell. The idea was further refined at Arden House, New York, in February 1992. Sir Crispin Tickell Interview with the author, August 10, 2012. See also: Engfeldt 2009, 171–176.

[78] The Malaysian lead on institutional questions was especially important because Malaysia had been a leading voice on the preparatory committee among developing countries. Lars-Göran Engfeldt Interview with the author, July 23, 2012.

[79] IISD 1992a.

[80] Engfeldt 2009, 176–177.

project in November 1990, a permanent fixture in the implementation of UN environmental activities. The GEF had developed a reputation for economy and expeditiousness, and had enjoyed success in mobilizing new funds.[81] Making the GEF permanent would combat the proliferation of single-issue multilateral funds, a source of fragmentation in UN environmental governance.[82] For developed countries, the GEF had the great virtue of being subject to the World Bank's competent management practices and maximized donor control over the use of funds.[83] Technical and financial criteria should, developed countries maintained, determine funding, not the political vicissitudes of one-country-one-vote fora.[84]

Developing countries argued that the undemocratic, opaque decision-making processes of the World Bank were an inappropriate follow-up mechanism for UNCED.[85] While the GEF was a joint project of the World Bank, UN Development Programme, and UNEP, the bank was clearly in the driver's seat.[86] Instead, China, together with the G77, advocated separate funds for the purposes of each environmental convention as well as the creation of a "Fund for the Promotion of Sustainable Development." Developed country contributions to these funds would be mandatory and should not be subject to any form of conditionality. The developing countries argued that mandatory funding mechanisms would enable greater predictability in development planning.[87]

The Rio Temporal Focal Point and Institutional Change

The North-South deadlock that characterized institutional bargaining following the release of the Brundtland report in 1987 persisted into the UNCED preparatory process. Developed and developing countries saw important benefits to change, but were reluctant to embrace fully the sustainable development formula for fear of signaling bargaining weakness. This section first demonstrates that UNCED meets the

81 Culpepper 1991, 15–16. 82 Brenton 1994, 208. 83 Brenton 1994, 188.
84 Werksman 1996, 153–154.
85 Preparatory Committee for the United Nations Conference on Environment and Development 1991.
86 Lars-Göran Engfeldt Interview with the author, July 23, 2012.
87 Preparatory Committee for the United Nations Conference on Environment and Development 1991.

definition of a Temporal Focal Point. It then proceeds to explain the relationship between the focal time frame and the realization of change in UN environment institutions in 1992. The *when* of change is, as we shall see, an important part of the *why* of change.

The Rio Conference as a Temporal Focal Point

The Rio Earth Summit meets the definition of a Temporal Focal Point. Temporal Focal Points are definite periods in time that stand out along the temporal continuum because of their conspicuousness and uniqueness. This subsection discusses each of these characteristics. UNCED was a definite period in time, running from June 3 to 14, 1992. There was no ambiguity over the timeline for acting within the ambit of the conference. The formal Rio preparatory period stretched from the passage of UN General Assembly resolution 44/228 in December 1989 to the conference proper in June 1992.[88] In addition to informal consultations throughout, four two-week formal preparatory committee meetings were scheduled: August 1990 in Nairobi; March 1991 in Geneva; August 1991 in Geneva; and March/April 1992 in New York. These four meetings gave a clear and identifiable structure to the Rio preparatory process.

UNCED was unique insofar as there were no similar temporal landmarks in the UN environment sphere at that time that could compete with it for the attention of actors. The May 1991 UNEP Governing Council sessions could not, for example, draw the eye of actors as a forum of comparable weight.[89] When member states of the UN General Assembly decided to urge governments to send representation at the head of state or head of government level, the exceptional standing of the conference was confirmed.[90] As one participant wrote during the preparatory process: "the Brazil Conference represents the best chance, perhaps the only chance, the world will have this decade to break the back of these global environmental issues."[91] Rio was regarded as a distinct and unparalleled opportunity for action.

The sense of uniqueness was bolstered by the twentieth anniversary of the landmark 1972 United Nations Conference on the Human

[88] United Nations General Assembly Resolution 44/228 1989.
[89] United Nations Environment Programme 1991.
[90] United Nations General Assembly Resolution 45/221 1990.
[91] Runnalls 1991, 4.

Environment and, to a lesser extent, the tenth anniversary of the 1982 UNEP Session of a Special Character. The coming anniversary of the Stockholm conference had been noted in 1987 by the Brundtland Commission as a potential moment to host a global conference, and was mentioned explicitly in the General Assembly resolution that launched the Rio conference.[92] Entrepreneurial actors, who aimed to regain the initiative in UN environmental politics, seized upon the legacy of the Stockholm conference to advance institutional reform.[93]

The conference was also highly conspicuous. Part of this conspicuousness had to do with broad public perceptions of environmental deterioration, fueled in part by a string of environmental disasters in the 1980s.[94] The destructive Exxon Valdez oil spill in March 1989 and the fact that the 1980s featured four of the five hottest years on record to that point (1981, 1983, 1987, and 1988) contributed to a popular sense of environmental crisis.[95] Public reactions to these events reached an "apocalyptic hue."[96] Reflecting the heightened public anxiety, *National Geographic* magazine devoted its special commemorative centennial edition in January 1988 to coverage of its environmental symposium, "Earth 88."[97] *Time* magazine abandoned its annual "Man of the Year" feature in 1989 in favor of a "Planet of the Year" edition. Popular concern moved environment, which had been a "fringy" issue before the 1980s, into the mainstream.[98] Thus, just as the UNCED process was getting underway, many observers were primed for a big event.

Just as importantly, the Earth Summit was one of the first major international gatherings since the end of the Cold War. While the breakdown of East-West confrontation did not directly create new opportunities for global environmental cooperation, it did contribute

[92] WCED 1987, 23; United Nations General Assembly Resolution 44/228 1989.

[93] Sir Crispin Tickell Interview with the author, August 10, 2012.

[94] Key events included the Bhopal chemical leak in 1984, the Chernobyl nuclear accident in 1986, the agrochemical leak in Basel in 1986, and the Exxon Valdez oil spill in 1989. For a discussion, see: Brenton 1994, 125–126.

[95] The Exxon Valdez, a California-bound oil tanker, foundered at Prince William Sound, Alaska, spilling more than eleven million gallons of crude oil into the ocean. Dramatic images of contaminated waters and devastated wildlife were common in the news media.

[96] Brenton 1994, 127, 166.

[97] National Geographic 1988.

[98] Shabecoff 1996, 114.

to making UNCED highly conspicuous.[99] Many governmental and nongovernmental actors looked to environment and development as potential avenues for renewed global cooperation through the United Nations. Environment was one area, many hoped, that the UN organization would make its mark in the post-Cold War years. The Earth Summit thus became a bellwether for the new era. Led again by Maurice Strong, UNCED's secretariat did much to boost the profile of the summit – traveling across the world to encourage a large turnout from world leaders and widespread media coverage for the Earth Summit.[100]

The conference, therefore, allowed actors to reach a temporal convergence of expectations. Unlike WCED, there was no ambiguity here. The Rio conference was a temporal landmark that was impossible to overlook. Prominent temporal signposts serve as a timing mechanism in international politics, enabling a convergence of expectation at certain points along the temporal continuum. Goal-oriented, strategic players time their policy interventions so as to maximize the likelihood that they will influence outcomes. Since actors invest scarce political and analytical resources in advancing reform proposals, they are reluctant to do so until they expect others to make a similar investment. They may, for example, hold off until a moment when they believe that other organizations with similar objectives will act. They may also wait until a time when they think that political actors are searching for allies or viable institutional alternatives. Much of this behavior is directly observable because actors seek to be observed.

In the Rio case, a TFP led to a dramatic increase in the investment of analytical resources devoted to exposing and solving the problems facing the global environment. In the lead-up to the Rio conference, we observe a proliferation of literature aimed at influencing world opinion. UNEP, vying for a leadership role in post-UNCED activities, produced a series of publications aimed at shaping policy debates and institutional alternatives. Most prominent among these was *The World Environment 1972–1992: Two Decades of Challenge*, an 884 page

99 Bernstein 2001, 86–87; Brenton 1994, 86–87.

100 Maurice Strong worked closely with Ted Turner to develop media coverage strategies for UNCED. In addition, the UNCED secretary-general worked with Jacques Cousteau, the celebrated French naturalist and filmmaker, to produce a documentary about the Earth Summit and its accomplishments. See: Strong 2001, 207.

follow-up to *The World Environment 1972–1982* that aimed to demonstrate UNEP's capacity for international environmental assessment.[101] UNEP and its enterprising Executive Director, Dr. Mostafa Tolba, also published at least three additional contributions for the Earth Summit.[102] Many other UN bodies, such as FAO, UNESCO, and UNCTAD, produced publications ahead of UNCED, seeking to highlight *their* important contributions to global environmental governance.[103] Intergovernmental bodies, such as the OECD and the World Bank, weighed in on key international problems and opportunities.[104] The Commonwealth of Nations authored a report on sustainable development, from the perspective of environmental protection, and the organization's former Secretary-General Sonny Ramphal authored a widely read follow-up to *Only One Earth*.[105] Not to be outdone, nongovernmental organizations and think tanks added to the growing literature surrounding the Earth Summit.[106] Such analyses also entered the popular realm. As the conference approached, Brazilian television stations, for example, produced extensive features on the global environment and famous artists became voluntary spokespersons for the cause. The Brazilian environment ministry even emerged as a popular brand featured in children's toys.[107] The popular conversation on environment was so widespread that even the Berenstain Bears weighed in during the focal phase.[108]

A number of seminars and prominent events were also held in this period. Pivotal conferences on institutional questions were held by the Aspen Institute and at Arden House in New York. In February 1992, UNEP convened workshops on environment and natural resource-accounting and on environment and economics.[109] The World Health Organization held special panels and commissioned reports in 1992 on health and environment, urbanization, food and agriculture, and energy.[110] In May 1992, the United States Senate Committee on Governmental Affairs held hearings on developing an energy-efficient

101 Holdgate et al. 1982; Tolba et al. 1992.
102 Tolba 1992.
103 For a sample of the set of publications from UN agencies that proliferated in the lead-up to FAO 1992; UNCTAD 1992; UNESCO 1992.
104 OECD 1992; World Bank 1992.
105 Commonwealth Secretariat 1991; Ramphal 1992.
106 Friends of the Earth 1992; SCOPE 1992; World Resources Institute et al. 1992.
107 Crocitti and Vallance 2012, 97.
108 Berenstain and Berenstain 1991.
109 UNEP (Hussein Abaza) 1992.
110 WHO 1992a, 1992b, 1992c, 1992d.

world.[111] Canada gathered "Stockholm veterans" for a December 1991 seminar in Meech Lake, Québec, to inform Canadian policy at UNCED.[112] As with the clustering of publications, the pre-Rio period gave rise to dozens of important seminars and events that treated various dimensions of the world environmental agenda. These discussions contributed to transforming the information environment facing actors in 1992, thus clarifying preferences nationally and internationally.

The investment of political resources in support of institutional change was subject to the same bursty, nonlinear pattern. Since international action in UN environment affairs required extensive coordination among hundreds of relevant actors, the timing of institutional initiatives was highly important. A poorly timed intervention could easily be overlooked or even backfire. UNCED's prominence made it the obvious time for actors to seek to build a supportive coalition for change. International projects that may have taken shape more gradually were paced to coincide with the Rio conference, so as to maximize the likelihood of agreement. This type of behavior was deliberate and strategic, as UNEP's former deputy executive director noted early in the Rio process:

> It is possible with foresight to use the preparatory process itself to produce tangible, completed action by the time of the conference. For example, work already underway on various draft conventions can be accelerated to be ready for early approval and signature [in Rio] ...[113]

Temporal-pacing of this sort was a large part of the Rio process. The decision to launch negotiations on a framework convention on climate change, pushed by the European Community and the Alliance of Small Island States, targeted an agreement for signature at UNCED.[114] Despite having just fifteen months to conclude negotiations on the most contentious of global conventions, states worked rapidly to reach agreement as a means of taking advantage of the coordinative impetus of the approaching summit.[115] The decision to pursue a convention on

111 United States Senate 1993.
112 Canadian Department of External Affairs and International Trade 1991.
113 Thacher 1991, 77.
114 Brenton 1994, 183–185; Haigh 1992, 247–248. The resolution that launched negotiations is: United Nations General Assembly Resolution 45/212 1990.
115 For a very fine account of these urgent talks, see: Engfeldt 2009, 185–186.

biological diversity in time for Rio was originally proposed by US diplomats, but – in a UNEP effort to find *a* role in the UNCED process – the initiative was hijacked by UNEP and its executive director.[116] Tolba received especially strong support from Sweden and Kenya in the biodiversity talks.[117] The US initiative to negotiate a world forest convention, which ultimately resulted only in a set of forest principles, is still another example of political initiatives launched to meet the Rio timeline.[118]

The approaching focal point also altered the political landscape within national capitals. Convergent expectations around a focal time frame led to major changes in domestic bureaucratic structures. In Brazil, for example, the planned conference, which was launched in no small part to deflect criticism of its environmental track record, made major bureaucratic investments in the tools of environmental policy and diplomacy. Within Itamaraty, the Brazilian foreign ministry, the government established a Division of the Environment and put it in the substantive lead for conference preparations. An interministerial commission was also created to clarify preferences over conference outcomes and prepare Brazil's national report for the conference. Ibama, the Brazilian environment ministry, was also shaped heavily by the approaching conference leading to several measures to bolster its weak bureaucratic standing.[119] In the United States, an interagency task force was established that was called the "Coordination Center" and the responsibilities of the Council on Environmental Quality (CEQ) and the State Department's Office of Oceans and International Environmental and Scientific Affairs were augmented considerably. As noted by Stephen Hopgood, however, these formal mechanisms were only a part of the intensive intrastate bargaining that was triggered by the approaching conference. These efforts grew more prominent by the day.[120] US preparations were shaped heavily by informal contacts and jockeying between key agencies, including the CEQ, the State Department, the Environmental Protection Agency (EPA), the Oval Office, the Council of Economic Advisors, the White

116 Hopgood 1998, 134, 169–170.

117 Engfeldt 2009, 186.

118 Hopgood 1998, 135.

119 For analysis of Brazilian policies toward the conference and its international environmental policies, see: Barros 2017; Corrêa doLago 2009; Hurrell 1991, 197–215.

120 Hopgood 1998, 141–142.

House Council on Competitiveness, and the Office of Management and Budget.

Taken together, the spike in the investment of scarce analytical and political assets in the lead-up to the Rio conference is striking. Important international actors were behaving *as if* there was a major nonlinearity or discontinuity in the passage of time. They paced negotiations on the climate change and biodiversity conventions to reach a conclusion in June 1992.[121] Agreement in September 1992 or March 1993, for example, was not seriously entertained. The heightened level of analytical activity in and around UN environmental institutions, moreover, contributed to an information-rich environment. Such analytical and political activities, both at the national and international levels, represented a very substantial increase in transaction costs. States accepted these costs because they expected others to make similar investments, devising institutional alternatives and bargaining accordingly. Such transaction cost-intensive investments were essential, moreover, to the clarification of preferences. In the United States, for example, intrastate bargaining helped to resolve intense internal policy debates, such as the one that unfolded between EPA administrator William K. Reilly and presidential Chief of Staff John Sununu.[122] 1992 thus became a landmark year along the temporal continuum, in large part, because all actors expected other actors to treat it as such. UNCED thus helped precipitate a temporal convergence of expectations.

The Impact of Convergent Expectations

The discussion above demonstrates that the 1992 Rio Earth Summit was a definite, unique, conspicuous moment in time that induced a temporal convergence of expectations among relevant actors. Concurrent expectations led to a sharp increase in the investment of political and analytical resources aimed at influencing UN environmental politics. This subsection focuses on the impact of the Rio TFP on the bargaining processes that resulted in a major transformation in core UN environmental institutions.

When states addressed institutional questions at UNCED and its preparatory meetings, there were important incentives to reach an

[121] Manulak 2020. [122] See: Hopgood 1998.

agreement. Exogenous conditions had changed, rendering current modes of cooperation suboptimal. The "react and retrofit" orientation of exiting institutions was poorly suited to address the underlying causes of global environmental decline. Through cooperation, developing countries could gain access to the technical and financial resources necessary to slow environmental degradation, which, they increasingly acknowledged, badly hindered their development efforts. Developed countries recognized a gradual worsening of the state of the world environment and sought to ameliorate pressing planetary problems through institutional change. Realizing sustainable development on a global basis offered potential gains to all sides.

A process-tracing analysis of the Rio negotiations reveals the important influence of TFPs in realizing change. First, the twentieth anniversary of the Stockholm conference and the approaching Earth Summit helped states to engage in bargaining without signaling weakness to the other side. Second, as UNCED neared, we observe a series of concessions, culminating at the end of the focal time frame. Actors sequenced their compromises, progressively eliminating the "square brackets" around contentious negotiating topics, until they were faced with an agreement that was treated as a final, *take-it-or-leave-it* offer. The urgency of the fast-approaching conference "focused the minds."[123] Although negotiations could have conceivably occurred following UNCED, actors behaved as if this was a practical impossibility. "As the Conference approached," one participant recalled, "the pressure on delegations increased."[124] Evidence of temporal-pacing is thus abundant.

Although informal consultations were held in summer 1991 and early 1992, formal negotiations on institutional questions did not pick up steam until the fourth preparatory committee meeting. The key breakthrough came on March 23, 1992, when Ismail Razali introduced a "non-paper" that, among other things, proposed the creation of a Sustainable Development Commission.[125] As the endpoint of a shared time frame neared, states behaved *as if* a firm deadline was in place and governed their behavior accordingly. The Stockholm anniversary allowed actors to point to factors other than their own keenness for agreement when entering into talks.

[123] Brenton 1994, 194. [124] Engfeldt 2009, 162. [125] IISD 1992a.

Over the course of the fourth preparatory committee meeting, the UNCED working group on legal and institutional questions bargained rapidly to narrow the range of disagreement over the design of institutional arrangements. To bridge the divide among delegations, an intense schedule of meetings – formal and informal – occurred over the final week of the fourth preparatory committee session.[126] A review of the excellent and highly detailed *Earth Negotiations Bulletin* – itself a significant nongovernmental political and analytical investment – documents the convergence of negotiating positions, as countries cooperatively managed risk by making dozens of reciprocal concessions over the course of negotiations. When the preparatory committee concluded its work on April 4, 1992, the working group was able to report "remarkable progress."[127]

Outstanding issues took "the political heat of Rio" to resolve.[128] The institutions negotiating group met on June 6, to address remaining questions regarding institutional arrangements. Key outstanding issues were: (a) whether the new commission would report "to" ECOSOC or "through" ECOSOC; (b) whether the commission should be designated a high-level body or not; and (c) whether G77 concerns about national reporting could be met.[129] These issues were resolved on June 8, when Razali introduced a revised text for governments' consideration.[130] What is important to note is that the desire to realize new cooperative possibilities forced actors to choose between the agreement that was attainable at Rio and the status quo. It effectively removed the "keep negotiating" option from consideration. Although most expressed concern about particular elements of the institutional alternative, it was Pareto-superior for all relative to extant arrangements. There was a Zone of Possible Agreement. Rather than continuing to

[126] On multiple occasions, the working group met twice in a day. On March 31, for example, a contact group met in the morning to work out key details and the full working group met from 8:30 p.m. until midnight, seeking to bridge the divide. See: IISD 1992b, 1992c.

[127] IISD 1992d.

[128] Brenton 1994, 218. In addition to the issues that remained unresolved at the end of the fourth preparatory committee session, several other important issues were introduced. The developing countries demanded that the CSD's monitoring provisions be loosened on the grounds that they were an invasion of national sovereignty. In addition, the CSD would concern itself with developed countries' follow-up on their commitments to increase financing for sustainable development.

[129] IISD 1992e. [130] IISD 1992f.

tweak the proposed settlement to find potential joint gains, actors took the agreement that they could get at Rio. Given the haste in which the post-UNCED institutional agreements were reached, this is an important fact. The desire of actors to reach a settlement before the close of conference, capitalizing on Rio's coordinative window, "outweighed bureaucratic caution" and enabled institutional change.[131]

The outcomes of the preparatory debates on institutional mechanisms for the political follow-up of UNCED were resolved through the creation of a high-level UN secretary-general advisory body on sustainable development. States also reached agreement on the creation of a functional Commission on Sustainable Development, under the UN Economic and Social Council, to review and monitor progress on the implementation of the UNCED action plan. The CSD was designated a "high level body," with the implication that its sessions should draw participation at the ministerial level and from central government agencies. The CSD had an extensive mandate in the follow-up on Agenda 21, UNCED's action plan. It would monitor progress in the UN system, make recommendations to the General Assembly, and analyze information provided by governments.[132] While states were not formally required to report to the commission, they would be encouraged strongly to do so.[133]

On financial mechanisms, the Global Environment Facility became the primary vehicle through which Agenda 21 follow-up would be funded. The GEF would also handle funding for the climate and biodiversity conventions. Developed countries acknowledged from early in the process that the GEF's informal, donor-driven governance

[131] Brenton 1994, 228. [132] Engfeldt 2009, 178–179.

[133] Developing countries worked at Rio to limit the reporting requirements of the CSD because they considered it an invasion on their sovereignty. At the same time, reporting requirements on the status of funding commitments were inserted. Some analysts – partially because the CSD did not turn into a strong forum – have argued that states' institutional achievements at Rio were disappointing. While some of the CSD's limitations were a product of deliberate institutional design choices, I argue that this assessment has been colored somewhat by subsequent history. The loss of momentum in global environmental governance after UNCED, including the inability of developed countries to meet their funding commitments and the unwillingness of governments to send high-level representation to the CSD limited the new body's efficacy.

structure would be unsuitable for post-UNCED follow-up.[134] As a result, there was a move to institute more participatory processes that increased the transparency of GEF decisions. National priorities would be taken into account in funding sustainable development projects, assuaging, to some extent, fears of green conditionality. The facility would strive toward universal membership and would be transparent and accountable to donor *and* recipient countries.[135] Developed countries, furthermore, reaffirmed their commitment to reach the 0.7 percent target for official development assistance as a percentage of gross national product.[136] Progress on reaching this target would be monitored by the CSD.

The institutional outcomes of the Rio conference linked environmental, social, and economic activities in the UN system. As Lars-Göran Engfeldt notes, post-UNCED institutional change "opened up the possibility for one international agenda for development rather than one for development and another for sustainable development."[137] In responding to change in underlying conditions, states adopted the "alternative agenda" of the Brundtland Commission. In a manner consistent with the recommendations of the Brundtland report, these measures were designed to provide badly needed coherence and direction to UN environmental institutions. What is remarkable about the institutional outcomes of the Rio conference is how closely they resembled the Brundtland Commission formula. Not only did states seek to institutionalize the logic of *Our Common Future*, they adopted many of the organizational modalities suggested by the commission. Although negotiations during the UNCED preparatory process were contentious, it is striking how narrow the range of debate was. Looking back, the Rio outcome was, according to WCED's chair, "enormously successful" from the perspective of implementing the recommendations of the commission she led.[138]

[134] Preparatory Committee for the United Nations Conference on Environment and Development 1991.
[135] For a discussion of GEF reform, see: Werksman 1996,154–155.
[136] Crucially, the United States was not bound by this commitment since it had not affirmed the target in the first place.
[137] Engfeldt 2009, 178.
[138] Gro Harlem Bruntland Interview with the author, August 17, 2021.

Conclusion

The 1987 report of the World Commission on Environment and Development, *Our Common Future*, formed the conceptual basis for institutional change at the 1992 Rio Earth Summit. The Brundtland Commission highlighted past changes in exogenous conditions. Significant shifts in the global economic setting rendered existing institutional arrangements suboptimal and created incentives for swift reform. Despite incentives to alter UN environmental institutions, states had been unable to act since the mid-1980s, effectively leaving gains on the table. The reason for the slow international response was the existence of severe temporal coordination problems. Coordination failure would prevent institutional change and badly harm cooperative players' negotiating position in future talks. States, North and South, had much to lose if they bargained cooperatively and others did not. The danger of exploitation in intergovernmental bargaining allowed the prevailing, risk-dominant equilibrium to persist.

The arrival of a Temporal Focal Point, the 1992 United Nations Conference on Environment and Development, provided the temporal landmark needed for actors to reach a temporal convergence of expectations. The highly conspicuous event contributed to a significant increase in the investment of analytical and political resources in UN environmental affairs. A wave of publications, conferences, seminars, and workshops proliferated, which added much to the recognition of cooperative potentialities, thereby transforming political and informational conditions. Although the range of disagreement among states concerning institutional change proposals was remarkably narrow, important differences needed to be ironed out. The presence of a TFP allowed actors to sequence a series of concessions, culminating in agreement at the Brazilian conference. The result was the creation of institutional machinery geared toward achieving sustainable development at the international level. New institutional bodies, such as the CSD and the reconstituted GEF, altered modes of cooperation and blended environmental, economic, and social considerations in development.

These findings support the theoretical approach set out in Chapter 2. In the presence of exogenous shifts, a temporal convergent of expectations is crucial to the realization of change. The slow pace of change seen through the mid-to-late 1980s, despite widespread environmental decline and institutional suboptimality, underlines the importance of temporal coordination problems in the realization of change. The emergence of prominent temporal signposts, furthermore, is crucial to resolving coordination dilemmas. The case study provides ample evidence to support the conclusion that convergent expectations play a decisive role in institutional change processes.

7 Post-UNCED UN Environmental Institutions

The preceding discussion has charted the political history of United Nations environmental affairs from the 1960s until 1992. The availability of archival documentation enables a dual approach to theory testing, including detailed assessment at both the outcome and process levels. Absent such archival documentation, it is not possible to reproduce the level of detailed, process-tracing for more recent phases of global environmental governance. It is, however, possible to analyze these phases at the level of comparative statics, assessing the presence or absence of the two independent variables specified in this book in producing the outcome of interest. Such an approach also has the virtue of policy-relevance and empirical interest.[1]

This chapter begins with a preliminary assessment of the post-United Nations Conference on Environment and Development (UNCED) period in UN environmental cooperation. This analysis is divided into three main sections. The first examines the period from 1993 to 2002, culminating in the 2002 World Summit on Sustainable Development, held in Johannesburg, South Africa. In this period, states sought to address important institutional ambiguities that resulted from institutional change at the 1992 Rio Earth Summit. Divergent expectations at the summit led to the persistence of the institutional status quo. The second section discusses the post-Johannesburg period, culminating in the 2012 United Nations Conference on Sustainable Development (UNCSD), held in Rio de Janeiro. The second Rio Earth Summit constituted a Temporal Focal Point in UN environmental politics and precipitated significant institutional change. The third section examines the post-2012 phase, providing a preliminary evaluation of the

[1] The inclusion of an additional chapter built around a number of shorter case studies as a means of increasing the number of observations is advocated in: King et al. 1994, 146.

current state of UN environmental institutions post-UNCSD. I conclude the chapter with a discussion of future prospects.

1993–2002: The Post-UNCED Hangover

Despite setting out an ambitious plan of action in the form of Agenda 21 and agreeing to substantial changes to cooperative structures that institutionalized sustainable development within the United Nations system, the decade following the 1992 Rio Earth Summit was defined by important challenges. In a way, the UN environment system was a victim of its own success in Rio. In the recollections of Elizabeth Dowdeswell, UNEP's new executive director, the post-UNCED period reminded UNEP staff of the old Zen saying, "after enlightenment, the laundry." For, she recalls, "we had all become enlightened through the Rio process and then all of a sudden we had to actually do something about it."[2] In the rush to adopt significant cooperative changes within a focal time frame, many significant institutional ambiguities remained to be ironed out following the Earth Summit. The temporal-pacing that had so profoundly affected bargaining processes in 1992 created a new set of institutional dilemmas that affected the implementation of Agenda 21 and the nature of environmental cooperation within the UN system more generally.

Perhaps the most important institutional challenge facing states was the relationship between the UN Environment Programme (UNEP), which had been marginalized in the Rio process, and the new Commission on Sustainable Development (CSD). What would be the division of labor between these two bodies and what were the comparative advantages of each? This challenge went beyond the usual growing pains facing new international fora. The creation of the CSD had very real implications for UNEP's cooperative mandate. As political backing of UNEP ebbed, financial support for it flagged.[3] Dowdeswell, sought to right the ship through badly needed internal reforms to the programme. These reforms encountered stiff resistance within the organization, however. Uncertainties arose also from residual ambiguities concerning the precise meaning and application of sustainable development in the UN context. As Dowdeswell recalled

[2] Elizabeth Dowdeswell Interview with the author, September 7, 2021.
[3] Ivanova 2012, 581–582.

in an interview: "The impact of Rio was not only substantive but raised questions about the future direction of UNEP. We asked, so where does environment fit within this ill-defined concept of sustainable development and how do we figure that out."[4] It was necessary, for instance, to establish what the implications of each of sustainable development's three pillars – economic, social, and environmental – would be for the distribution of responsibilities between UNEP and the CSD, and for intergovernmental priority-setting more generally.

The proliferation of multilateral environmental agreements (MEAs), which continued in the 1990s, had further implications for broader UN environmental relations. Emerging out of the Rio process was the Convention on Biological Diversity, headquartered in Montréal, Canada, the UN Framework Convention on Climate Change (UNFCCC), and, eventually, the UN Convention to Combat Desertification, both based in Bonn, Germany. Combined with the traditional hesitancy of UN specialized agencies and funds to be coordinated by UNEP – or any other central UN body for that matter – the growth of MEAs deepened fragmentation in UN environmental institutions and diminished overall coherence.[5] "Apparently," recalls Dowdeswell, "there had been little thought given to or agreement reached about implementation ... We talked about reducing fragmentation but 'don't you dare try and coordinate me ...'" This system involved a particularly high degree of competition for funds and for autonomous implementation mandates.[6] Importantly, fragmentation deepened gradually, mounting with each new convention and intergovernmental initiative. As a consequence, the post-UNCED system of environmental cooperation came to deliver less than the sum of its parts.

These institutional challenges were exacerbated by deepening North/South polarization in this period. Economic recession in the 1990s and the socio-economic aftereffects of the end of the Cold War helped to produce a period of restraint in the financing of many international initiatives, including sharp reductions in funding available for official development assistance. Miserly development assistance allocations

[4] Elizabeth Dowdeswell Interview with the author, September 7, 2021.

[5] Some environmental conventions, such as the Montréal Protocol, the Convention on Biological Diversity, as well as the Basel, Stockholm, and Rotterdam Conventions, are at least in part administered by UNEP.

[6] Elizabeth Dowdeswell Interview with the author, September 7, 2021.

were seen in many developing countries as a breaking of the Rio bargain, a violation of the UNCED sustainable development deal. This limited the willingness of many developing countries to cooperate on the green priorities of developed countries and contributed to worsening institutional dysfunction. At the same time, action on many pressing environmental questions was being realized increasingly through the MEAs. This reduced somewhat the urgency associated with addressing institutional limitations and the mandate ambiguities of UNEP and the CSD.

Despite institutional problems, states struggled to regain the coordinative impetus of the Rio process. Sustainable development had been implemented in an incomplete manner, leaving cooperative incentives unrealized. Governments could not bring themselves to, in effect, solidify the beneficial institutional change arranged in Brazil. Numerous ongoing efforts to tackle institutional suboptimality met with disappointing results, furthermore. This process involved the negotiation of the Nairobi declaration in 1997, the creation of a task force under former German Environment Minister Klaus Töpfer, and the Malmö declaration in 2000. Pursuant to UN General Assembly Resolution 53/242, states sought to strengthen system-level coordination in the environmental and human settlements fields through the creation of the Global Ministerial Environment Forum of UNEP and the Environmental Management Group.[7]

All of these efforts helped incrementally to clarify the mandate of UNEP, the CSD, and the other relevant organs of UN environmental governance. Each process pointed to the need for large-scale change in UN environmental cooperation to address post-Rio challenges and bring institutions into line with ever-changing global conditions. These processes highlighted duplication, governance gaps, and coherence problems.[8] To address these challenges and capture available incentives to improve overall system coherence, international actors advocated a variety of measures, including expanding UNEP's Governing Council to encompass universal membership, upgrading UNEP from a subsidiary body of the UN General Assembly into a specialized agency, or transforming the moribund UN Trusteeship Council into a vehicle for advancing sustainable development. While

[7] United Nations General Assembly Resolution 53/242 1999.
[8] Andresen 2007.

many ideas floated about, limited action was taken by governments through the 1990s. These ongoing processes represented modest political and analytical investments in change processes at the intergovernmental level. What was lacking beneath the surface of intergovernmental discussion, however, was a corresponding national-level engagement with the subject on a wide international basis. Many governments did not clarify their preferences among institutional alternatives, for example. These discussions thus lacked the urgency seen prior to the UN Conference on the Human Environment and UNCED, failing to provoke a more decisive shift in bargaining strategies.

As the new millennium approached, calls for reform were directed increasingly toward preparations for a planned tenth anniversary of the Rio conference. The World Summit on Sustainable Development (WSSD), scheduled for August/September 2002 in Johannesburg, aimed to recapture the coordinative impetus of past UN conferences. Promisingly for reformers, the UN General Assembly Resolution launching WSSD focused on implementation of Agenda 21 and pointed explicitly to institutional questions as a focus for the conference.[9] The call on governments to participate at the summit level was also encouraging for reform-minded actors, since it was an indication of high-level political attention.

Despite these favorable circumstances and incentives to address underperforming institutional mechanisms, WSSD was characterized by a temporal divergence, rather than a convergence, of expectations. Unlike the period from 1993 to 2001, the Johannesburg summit set out a clear, discrete time frame for action: August/September 2002. Like previous conferences, preparatory meetings were arranged to build a substantive framework for the conference.[10] Yet, unlike UNCED or the 1972 UNCHE, the South African meeting was neither unique nor conspicuous. Diminishing WSSD's uniqueness was the occurrence of other, closely related conferences in this period that competed for the limited attention of global actors. Most importantly, negotiations over the Monterrey consensus at the March 2002 United Nations Conference on Financing for Development had very important

[9] United Nations General Assembly Resolution 55/199 2001.

[10] Four preparatory committees were organized for WSSD. The first three were held in New York (April/May 2001, January/February 2002, and March/April 2002) and the fourth was held in Bali, Indonesia (May/June 2002).

consequences for broader North/South relations in the UN context. Similarly, the World Trade Organization's Doha Development Round, which opened in November 2001, affected the sustainable development agenda and impacted fundamentally the uniqueness of WSSD. Indeed, some states went so far as to advocate the explicit subordination of WSSD agreements to the Doha outcomes.[11] These and other related events contributed to "conference fatigue," making Johannesburg appear as just one gathering in a series of similar meetings.[12]

The conspicuousness of the Johannesburg summit was affected by events outside the control of conference organizers and by choices made early in the preparatory process. The international shock delivered by the September 11, 2001 terrorist attacks in the United States, less than a year before WSSD, had a major impact on the broader international setting. The subsequent international response to the 9/11 attacks and the ensuing UN-sanctioned war in Afghanistan became an intense focus for governments – especially the United States and its allies – and consumed much of the oxygen in New York, where the WSSD preparations were taking place. Indeed, the attacks occurred just as the conference preparatory process should have been hitting its stride. These events left little room on the global agenda and little available bandwidth for big institutional investments outside of the counter-terrorism space. Funding that might have helped to support the sustainable development agenda was redirected to fight a "war on terror."

In addition, the decision was taken early in preparatory process that, unlike UNCHE or UNCED, WSSD would be organized within the UN's existing institutional framework, rather than through a dedicated conference secretariat. According to Lars-Göran Engfeldt, who served on the conference bureau, this setup detracted from the Johannesburg summit's visibility.[13] In a difficult and crowded international setting, particularly post 9/11, UN Secretary-General Kofi Annan also hesitated to prioritize the coming summit in his public engagements, focusing instead on other, competing priorities.[14] Consequently, the South African meeting was not spotlighted in the energetic way that previous meetings had been. The conference bureau also arguably lacked the

[11] Dodds et al. 2012, 113. [12] Engfeldt 2009, 245; Wapner 2003, 2.
[13] Engfeldt 2009, 253. [14] Dodds et al. 2012, 108; Engfeldt 2009, 256–257.

panache and keen political acumen of an entrepreneurial figure, such as Maurice Strong. The South African meeting was not a Temporal Focal Point.

While the conference drew significant interest and attention in 2002, it was simply not the UN environmental *phenomenon* seen in previous decades. The emergent properties characteristic of convergent expectations did not set in. The agenda simply did not cohere into anything beyond an obligatory, largely *pro forma* UN mega-conference.[15] Although the summit featured comparable attendance from governmental and nongovernmental organizations to UNCED, accredited media participation was less than half of what it had been ten years earlier. This reflected the limited profile of the Johannesburg summit relative to its Brazilian predecessor. While extensive, furthermore, large nongovernmental organizations, such as the International Union for Conservation of Nature and the Worldwide Fund for Nature, as well as the scientific community, faced uncertainty concerning avenues for influence in the South African meeting.[16] Stakeholder engagement was limited greatly in the run-up to Johannesburg, with limited national consultations and regional analysis conducted.[17] Consideration of institutional questions at WSSD, observes Maria Ivanova "had little in terms of an analytical foundation and failed to deliver a substantial, actionable proposal that responded to the political possibilities at the time." "No analyses," for instance, "dealt systematically with the question of why a World Environment Organization had not been established and why UNEP was not fulfilling the role that so many had hoped it would."[18] Diminished expectations for WSSD were a product – and a cause – of defensive posturing

[15] Paul Wapner provides an evocative summary of his observations from WSSD:

> At times, sadly, it seemed like most people came more to say their piece than to listen, learn or dialogue with each other. In plenary sessions, governmental ministers made prepared statements that were largely unrelated to the specific topics at hand. NGOs held panels that often failed to speak to each other or generate a larger vision for the future of environmental protection and world development. Even ordinary observers appeared lost in the shuffle as they shuttled frenetically from one event to another, hoping to catch a well-known speaker or see a dazzling exhibit. As should be expected at any such gathering, there was little time to absorb what was being said. Indeed, one got the sense that, while many words were being spoken, few were listening.

See: Wapner 2003, 2.

[16] Engfeldt 2009, 252. [17] Dodds et al. 2012, 113. [18] Ivanova 2021, 207.

among governments in institutional bargaining. The compromise and sustained exploration of the diplomatically possible seen in other global environmental talks did not materialize in the run-up to Johannesburg. The political and analytical investments made in the context of past meetings did not materialize in the run-up to WSSD.

Johannesburg, in the words of one scholar who observed the meetings closely at the time, was a "flop."[19] The results of the summit were, particularly for those anticipating institutional change, disappointing. The Johannesburg declaration and plan of implementation were underwhelming, doing little to build off of the Nairobi and Malmö meetings or address the institutional defects identified through the 1990s. States failed to capitalize on available change incentives and a temporal divergence of expectations reflected coordination failure. The institutional status quo largely prevailed. While there was attention to the CSD, including on its role and function within the UN system, the Johannesburg Plan of Implementation did not mention "environmental governance."[20]

2003–2012: The United Nations Conference on Sustainable Development, Rio+20: An Unexpected Success

Following the Johannesburg summit, many of the institutional limitations – and the incentives available to address them – persisted further into the first decade of the new millennium. Though problems were widely recognized, governments had not been able to adapt the institutional structures of UN environmental cooperation to prevailing international conditions. As a consequence, many problems of the world environment, acknowledged by governments decades earlier, were left un(der)addressed as preparations began for the 2012 UNCSD.[21] After twenty years, many elements of Agenda 21, the UNCED action plan, were not implemented.[22] UN environmental institutions continued to be afflicted with a high degree of duplication, overlap, and numerous governance gaps. UN agencies continued to operate with minimal coordination, failing to fulfill the vision of

[19] Author's email correspondence with Peter M. Haas, December 25, 2020.
[20] Dodds et al. 2012, 160. [21] Ivanova 2012, 573–575.
[22] Stakeholder Forum 2012.

institutional designers in Stockholm and Rio. The MEAs continued to foster cooperation in areas of more limited institutional scope, while contributing also to fragmentation and a degree of incoherence in the UN environmental sphere.

While the decisions taken at UNCED in 1992 had sought to bring environmental, social, and economic dimensions into closer alignment, as proposed by the World Commission on Environment and Development, this sustainable development approach had not been effectively implemented. Rather than realizing a coherent overall vision, there were inconsistencies in government actions in the UN context between the three pillars of sustainable development. Indeed, as Alex Evans and David Steven note, the Millennium Development Goals (MDGs), adopted as the sustainable development agenda should have been taking hold within the UN system, "left out key issues in both the economic sphere (e.g. growth, jobs, private sector, inequality) and the environment (e.g. climate change, water, biodiversity)."[23] As a result, in the early 2000s, there were two distinct agendas related to development: sustainable development and a more traditional form of development largely unencumbered by environmental imperatives. Incoherence was exacerbated as the CSD's review functions related to funding implementation of Agenda 21 were effectively dropped from its mandate, leading many development ministers to stop attending annual meetings.[24]

The residual mandate confusion associated with the decision in 1992 to launch the CSD continued to hamper cooperation within UNEP and elsewhere within the UN system. The level of governmental representation within the CSD had, moreover, fallen off noticeably. Intended to bring high-level attention to issues of sustainable development, thereby facilitating coordination, both across the UN system and from a national whole-of-government perspective, the CSD increasingly attracted delegations of bureaucrats from national environment ministries, rather than the ministers themselves. This was far from the CSD's intended audience of foreign, trade, development, and finance ministers. Put simply, the CSD was not serving its intended purpose and was increasingly regarded as a global "talk shop." As preparations for Rio+20 were coming into focus in spring 2011, states failed to reach agreement within the CSD on a chair's text, the second such

[23] Evans and Steven 2012, 3. [24] Dodds et al. 2012, 126–128.

failure in a four-year period. The commission was losing credibility and support in donor countries.[25]

The need to reform UNEP had been acknowledged explicitly by governments at the ministerial level in 2000 through the Malmö declaration.[26] Many of the institutional solutions to these problems had long been in circulation. The proposed transformation of UNEP from a subsidiary body of the UN General Assembly to a specialized agency, a proposal backed by both the European Union and the African Union, had been first proposed in the context of the 1972 Stockholm conference. At WSSD, the idea was endorsed by French President, Jacques Chirac.[27] The specialized agency concept enjoyed renewed prominence after Germany, Brazil, Singapore, and South Africa voiced strong support for the idea in 1997.[28] Other actors advocated universal membership for the UNEP Governing Council.

The necessity of enhancing the coherence of UN environmental institutions was acknowledged again by governments in 2005 and picked up early in 2006 in a series of informal consultations.[29] In 2008, the UN's Joint Inspection Unit noted the impact of the current fragmentary order on international environmental governance.[30] Such efforts were reinforced in 2009 at the UNEP Governing Council session and by the circulation of the "Nairobi-Helsinki" outcome document in 2010.[31] Many of these assessments, in effect, recycled conclusions discussed previously and, while demonstrating growing political awareness of institutional challenges, made few analytical or conceptual breakthroughs. What was needed, moreover, was not more international recognition of the challenges facing UNEP and the CSD. Instead, key governments had to make the necessary analytical investments at the national level to clarify their preferences in light of the prevailing international environment. In the context of divergent temporal expectations, few governments were willing to make the substantial investments required to do so. As a consequence, when these matters were considered for action, thus moving beyond declarations and exhortations, leading international actors were unwilling to commit definitively to available institutional alternatives. The divergent expectations that came to define WSSD thus persisted.

[25] UN DESA 2011, 17. [26] United Nations Environment Programme 2000.
[27] Chirac 2002. [28] For a discussion, see: Biermann 2000.
[29] Berruga and Maurer 2007, 16–25. [30] Inomata 2008.
[31] United Nations Environment Programme 2010.

The prognosis for tackling these challenges worsened further in the wake of the May 2009 Copenhagen Climate Conference. The failure of states to achieve a substantive new climate change agreement, despite the significant efforts of UN Secretary-General, Ban Ki-moon, to hype the event and implore states to "seal the deal," led many to question the efficacy of UN multilateral diplomacy in the early twenty-first century. Despite a highly visible event featuring summit-level participation, the inability to reach an agreement led the Danish conference presidency to force-feed the international community an agreement that was widely viewed as lacking transparency and legitimacy.[32] The acrimonious conclusion, note Macharia Kamau et al., created mistrust in UN diplomacy and haunted subsequent environmental cooperation.[33] In the recollections of Christiana Figueres, the UNFCCC executive secretary appointed in the aftermath of the Copenhagen meeting, and Tom Rivett-Carnac: "Copenhagen was nothing short of a disaster," yielding a "weak, inadequate accord that was politically unacceptable and legally irrelevant."[34] The proposed accord was, as exclaimed by the Venezuelan representative at the meeting, a "coup d'état against the United Nations."[35] Prospects for convergent expectations in the wider UN environment sphere dipped accordingly.

This reality changed with the emergence of a Temporal Focal Point: the 2012 UNCSD. Held in Rio de Janeiro, Brazil, twenty years following the original Rio summit, the conference was widely referred to as "Rio+20." The meeting, which took place in the context of a crisis in the Eurozone and a US presidential election, was organized in the face of deep pessimism. Many observers predicted underwhelming

[32] The failure of the Copenhagen Climate Conference is highly interesting for the model presented in this book, since the event clearly constituted a Temporal Focal Point. Seeking a revised climate accord, the UN and the Danish government worked to increase the conspicuousness of the event. There was widespread public interest in the meeting, leader-level participation, and – in the context of the UNFCCC – an urgent need for agreement. Yet, in 2009, states lacked a bargaining range. The efforts to find available joint gains, despite the focal event, produced a highly visible failure. Somewhat like the 1982 Nairobi conference result, the Copenhagen story is therefore illustrative of perils associated with seeking to crystallize a TFP in the absence of available joint gains.

[33] Kamau et al. 2018, 31–33. For a flavor of the acrimonious outcome of the conference, see: Ban 2021, 272–275.

[34] Figueres and Rivett-Carnac 2020, 47–48.

[35] Black 2009.

results, including on the now long-standing challenges facing United Nations environmental institutions.[36] This was especially so as negotiations on green economy cooperation did not advance for lack of an available bargaining range. Despite this pessimism, the conference came to provide a focus for governmental and nongovernmental actors. A total of 4,000 side events took place in Rio alongside the main conference. As many as 600 voluntary commitments were made by businesses and civil society groups, including the mobilization of over US$513 billion in financial commitments. Numerous declarations and principles were coordinated by non-state actors.[37] Comparing Rio +20 to the Johannesburg summit, Pamela S. Chasek and David L. Downie observe that the meeting "received far more attention" and "enjoyed a far more productive atmosphere."[38]

As it neared, UNCSD emerged as a conspicuous phase in UN environmental politics. Part of its conspicuousness owed to a concerted effort by the Brazilian government late in the process to increase the visibility and profile of the meeting. At stake, in the words of the International Institute for Sustainable Development's *Earth Negotiations Bulletin*, was the legacy of UNCED and the "spirit of Rio" itself.[39] Indeed, reflecting on the first Rio conference as well, a productive Rio+20 outcome was regarded by many as integral to "saving [the legacy of] two summits." With the UN in the throes of a post-Copenhagen credibility crisis, moreover, success at UNCSD was seen as essential to preserving the "reputation of multilateralism itself."[40] This spirit contributed a sense of urgency, particularly in the latter stages of the preparatory process. As states gathered in Brazil, the Rio spirit was manifested physically as "elders" of the 1992 Earth Summit walked the corridors and urged action that measured up to the legacy of the first Rio Earth Summit

While, to some extent, the broader UN environmental enterprise was by this time overshadowed politically by annual Conferences of the Parties of the UNFCCC, the most high-profile global environmental forum in this period, Rio+20 came to be seen as a unique opportunity to tackle several long-standing and persistent governance

[36] Rocard 2012. [37] Ivanova 2013a, 9–10.
[38] Chasek and Downie 2021, 16.
[39] International Institute for Sustainable Development Reporting Services 2012, 19.
[40] Kamau et al. 2018, 44–45.

challenges. As noted above, ongoing treatments of these challenges had not yielded coordinated and concerted action by governments. The dilemma was summarized by leading observers in Rio: "If not now, when? If not here, where?"[41] Visible failures, including in Copenhagen, added further impetus as states sought to show that UN-based multilateralism could still succeed. There was thus a *put up, or shut up* quality to institutional discussions at UNCSD that led international actors to treat Rio+20 as a unique moment along the temporal continuum. The conference, therefore, constituted a Temporal Focal Point in United Nations environment politics.

In the run-up to the June 2012 meeting, momentum grew toward the reform of UNEP. Among the principal institutional alternatives on the table were: the transformation of UNEP into a specialized agency or a suite of measures that increased the programme's capacity in its current form as a subsidiary body of the UN General Assembly. A focal time frame enabled the push required to focus the minds of leading actors, thus stimulating the political and analytical investments required to act on persistent institutional suboptimality.[42] For the UN secretary-general at the time, the Rio process served as "the culmination of a two-year immersion in humankind's impact on the environment."[43] Negotiations on institutional alternatives were halted as governments arrived in Rio. Using its prerogative as chair, Brazil produced and circulated what was effectively a *take-it-or-leave-it* draft of conference decisions based on negotiations during the conference preparatory process. Ultimately, governments agreed on the expansion of the UNEP Governing Council to encompass universal membership. While it would not be upgraded to a specialized agency within the UN system, universal membership enhanced the programme's legitimacy and was a characteristic of specialized UN agencies. This step, it was hoped, would strengthen UNEP's hand vis-à-vis the MEAs and ameliorate coherence problems. Measures were also adopted to provide the programme with more extensive, stable, and predictable funding.

Two other important institutional initiatives emerged in the context of the second Rio Temporal Focal Point. First, states elected to take the rare decision to put an institutional end to an international forum, killing the CSD just two decades after it was created. In its place,

[41] IISD 2012a, 20. [42] Ivanova 2012, 223. [43] Ban 2021, 254–255.

governments agreed to launch a High-level Political Forum (HLPF) on Sustainable Development that would carry out many of the original aims of the CSD. Discussions on what became the HLPF first emerged at a high-level dialogue on the institutional foundations of sustainable development in Solo, Indonesia, over a year into the UNCSD preparatory process. The background paper for the meeting, commissioned by the UN's Department of Economic and Social Affairs and penned by Stakeholder Forum's Felix Dodds, proposed five main institutional options at varying levels of ambition.[44] States discussed possibilities for reforming the CSD and UN's Economic and Social Council (ECOSOC) to meet the challenge. Among the most provocative of proposals was the creation of a Sustainable Development Council (SDC), modeled after the UN's Human Rights Council.[45]

Discussions reached a turning point in spring 2012 during a second round of "informal informal" consultations in New York. There, G77/China floated the idea of a High Level Political Forum on Sustainable Development. Despite some concern from African countries, led by Kenya, the proposal enjoyed preliminary support as governments urgently searched for an improvement over the CSD in time for Rio +20.[46] The HLPF concept had the virtue of being flexible and less institutionalized than the SDC proposal, yet more substantive than tweaking the CSD. Momentum grew when the US delegation, which had consistently opposed the SDC proposal, recognized the HLPF concept as "very useful" at the third and final "informal informal" consultation, just days before the opening of the Rio conference.[47] During the third UNCSD preparatory committee meeting and at the Rio conference itself, discussion focused on the forum's proposed functions and its relationship to other UN organs. As a result, the HLPF proposal, emerging just months prior to the Rio conference, was endorsed by governments and was included in the draft conference outcome document.

The HLPF would facilitate representation at the ministerial and head of government level, providing "leadership, guidance, and recommendations" for sustainable development. This representation included

[44] UN DESA 2011.

[45] For a discussion of the concept, developed by the well-connected nongovernmental organization, Stakeholder Forum, see: Strandaneas 2012.

[46] International Institute for Sustainable Development 2012a.

[47] International Institute for Sustainable Development 2012b, 7.

annual meetings under the auspices of ECOSOC and quadrennial summit-level meetings at the General Assembly. This latter feature targeted higher-level engagement than was typically of the CSD. For Steven Bernstein, the HLPF, which was envisioned as a stronger body than the one it replaced, had "enormous potential" to serve as a vehicle for orchestration in global environmental governance.[48] Yet, the speed around which the HLPF negotiations progressed, indicative of temporal-pacing, left numerous unresolved questions that had the potential to reproduce some of the same challenges that had plagued the CSD.[49]

In addition, Colombia introduced and built a supportive coalition in favor of what came to be known as the Sustainable Development Goals (SDGs). The SDGs would replace the Millennium Development Goals, which were set to expire in 2015. The SDGs held the potential to unify the bifurcated development agenda that had characterized post-UNCED cooperative efforts. Importantly, unlike the MDGs, the SDGs would be universally applicable to all states: developed and developing. There would not be one agenda for developing countries that concentrated on development, and another, for the developed countries, that focused on the provision of development assistance. The SDGs would target a global vision, helping to bring political attention, public engagement, and global action to a clear set of priorities.

The SDG concept – first outlined in a January 2011 briefing paper – had been the subject of corridor discussions for months, with Colombian diplomats seeking to "propose and position" the idea to key constituencies.[50] These constituencies included governments, North and South, as well as non-state actors. The SDG proposal was formally introduced at the July 2011 high-level meeting on the Institutional Framework for Sustainable Development in Solo, Indonesia. In the second half of 2011, Colombian officials pitched the proposal at diverse fora, including consultations hosted by Brazil, meetings of the Economic Commission for Latin America and the Caribbean, and the UN General Assembly. Although the idea garnered strong support from, among others, Guatemala and Peru, the proposal was met with derision and resistance from other actors. Many

[48] Bernstein 2013, 19. [49] Ivanova 2012, 224.
[50] Paula Caballero Interview with the author, July 30, 2021.

countries from the G77 and China Group, including India, initially opposed new metrics for quantifying any resulting goals, since they feared that this could result in aid conditionality.[51] Divisions were not moreover primarily North/South in nature, but were instead often concentrated *within* governments themselves. While the SDGs came to enjoy support in many foreign and environment ministries, many of those in charge of international development and cooperation agencies opposed the concept.[52] Influential development professionals argued that the proposed SDGs would undermine the MDGs, conferring an institutional "lame duck" status and compromising their poverty focus. In the words of Paula Caballero, the chief architect of the SDG concept within Colombia's foreign ministry, "[n]ot all negotiators appreciated the linkages between functional, resilient natural systems and long-term human well-being."[53] Many had hoped that the MDGs would be updated, renewed, and then promptly rolled-over into the post-2015 period.

A landmark in this first phase of negotiations was confirmation in November 2011 from the UNCSD secretariat that the SDG concept would be included within the "Zero Draft" of the Rio outcome document. This draft, circulated by the co-chairs of the UNCSD preparatory committee on January 10, 2012, would form the basis for the Rio +20 negotiations. While the idea had generated some initial support by late 2011, including among civil society organizations, justifying its inclusion in the draft, there was much transaction cost-heavy work remaining. In November 2011 and January 2012, Colombia hosted a pair of international consultations that sought to inform substantive consideration of the SDG concept. Both sessions, which were held, respectively, in Bogotá and Tarrytown, New York, were well-attended and generated significant interest as states prepared for UNCSD. These consultations, in the recollections of Caballero, aided her in "mapping the negotiation process and the political economy around the SDGs."[54] Critically, they also helped participants from across the

[51] Evans and Steven 2012, 1. Indian resistance was overcome via a visit by Caballero and Colombian Vice Minister Patti Londoño to New Delhi, seeking and obtaining support for the SDG concept. This support proved crucial as UNCSD neared.

[52] Paula Caballero Interview with the author, July 30, 2021.

[53] Caballero 2016.

[54] Paula Caballero Interview with the author, July 30, 2021.

geopolitical spectrum to think through and understand the concept, a key development given that discussions surrounding the SDGs were highly charged from the outset. The cross-cutting dynamics were "frightfully complex," with numerous interest cleavages coming into sharper focus.[55] Equally important, these discussions allowed delegates to return to their capitals in a position to clarify their national preferences in relation to the SDG proposal.

In addition to the consultations, Colombia continued to generate concept papers to help delegates understand the contours of the proposal and to frame and guide the negotiations. These concept papers were issued periodically during talks to help steer the political process.[56] Fundamental to these papers, and Colombian efforts more generally, was the desire to produce a comprehensive, universal agenda that would unify all actors in advancing sustainable development on a global scale.[57] On the basis of the Zero Draft, the formal negotiation process to finalize the outcome document for UNCSD kicked off in early 2012 with a suite of monthly sessions. Progress was painfully slow. As the June 2012 summit neared, many feared that time was running out. The negotiation text spiraled: the Zero Draft ballooned from a relatively tidy 128 paragraphs in January to over 400 in May, of which only 21 had been agreed *ad referendum*.[58] An additional negotiation session had to be scheduled.

Colombia spearheaded an effort to determine the post-Rio+20 process for defining the SDGs. Negotiation on what became known as the Open Working Group (OWG), were, recalled Caballero, "almost harder than on the SDGs" themselves.[59] States recognized that the process of specifying the goals would be highly important for their eventual character. Colombia, and a few other governments, including Mexico, held out until the end for a science-based, technical, inclusive working group that would define in a transparent and informed way the goals. This ran into strong opposition from many governments that

[55] Paula Caballero Interview with the author, July 30, 2021.

[56] Four concept papers were issued: (i) by Colombia only; (ii) endorsed by Guatemala; (iii) endorsed by both Guatemala and Peru; and (iv) endorsed by Peru and the United Arab Emirates (elections in Guatemala impeded their endorsement of the final paper, as the new government was in the process of defining positions and priorities).

[57] Paula Caballero Interview with the author, July 30, 2021.

[58] Caballero 2016.

[59] Paula Caballero Interview with the author, July 30, 2021.

either wanted a formal political process, such as that for Rio+20, or an effort led by the secretary general, such as a high-level panel.[60]

Through these intensive discussions, the Colombian proposal increasingly came to be regarded as a concrete deliverable from the Rio+20 process. The Brazilian Minister of Environment, Isabella Teixeira, backed the proposal strongly. As governments understood the scope and intent of the proposal, clarifying their preferences, many started to come around. India, while initially hesitant, became a critical supporter through the intergovernmental negotiation process. The SDGs generated growing political and financial support from a number of governments, including Bhutan, Mexico, the Netherlands, and Norway, as well as non-state actors. The World Resources Institute, a powerful Washington-based think tank, became a key supporter of the Colombian initiative.

In the context of a temporal convergence of expectations, the SDG proposal emerged as a means of realizing available change incentives in UN environmental and sustainable development governance. The rapid pace of discussions, entailing a sudden willingness to incur a significant increase in transaction costs, is indicative of temporal-pacing. Actors made very substantial political and analytical investments to evaluate the SDG concept in the context of current international realities. As the Rio+20 TFP neared, governments worked fervently to finalize the description of the concept, emphasizing the universality of the goals and the process that would enable their development. This text formed the basis for relevant sections of the General Assembly resolution formalizing the Rio outcome.[61] Moreover, states' willingness to endorse the SDG proposal at UNCSD, just a year and a half after the idea was first conceived, is illustrative of the power of Temporal Focal Points in global politics. States committed to the Colombian proposal, reflecting the desire of governments to seize significant opportunities for change in patterns of cooperation in relation to the focal time frame. If discussions were to wait until all of the specifics were finalized at the working level, actors feared that the initiative would eventually founder.[62] Even sympathetic observers were somewhat bewildered by the pace of the SDG

[60] Paula Caballero Interview with the author, July 30, 2021.
[61] UNGAR 66/288 2012. [62] Caballero 2016.

negotiation process, sounding a note of caution and urging governments to "take a gradual approach" and not "try to do everything at once in Rio."[63] International organizations are rarely accused of moving too quickly, underlining the remarkable progress possible when Temporal Focal Points emerge.

After intense negotiation, including discussions held just days before heads of state and government were scheduled to arrive in Rio, a proposal was worked out on a formula to articulate the goals. Agreement was reached on a thirty-member OWG that would produce the SDGs for the consideration of the UN General Assembly. Once constituted, the group would have a "narrow window," opened by UNCSD, to "ensure the smooth passage of short, meaningful and inspirational goals that [could] capture the public imagination."[64] Failure to achieve this could result in the type of institutional stasis seen in the two decades between UNCED and UNCSD. The thirty-member OWG proposal was included – and adopted – in Brazil's draft outcome document.[65]

As a result, 2012 came to produce significant non-incremental change in UN environmental institutions. The strengthening of UNEP, the discontinuation of the CSD, the launch of the HLPF, and the Sustainable Development Goals transformed current modes of environmental cooperation. Some of the measures, such as universal membership for UNEP's intergovernmental body, had been under discussion for more than a decade. Yet states could not bring themselves to take definitive action to realize available change incentives. Divergent temporal expectations explain the persistence of the institutional status quo. In other cases, such as the SDGs, progress was achieved rapidly as a consequence of substantial political and analytical investments made in the context of a Temporal Focal Point. The focal time frame helped to generate convergent expectations, enabling the mobilization of latent coalitions that proved instrumental to realizing institutional change in Rio. Governments clarified their preferences swiftly and acted to advance their interests in altering institutional arrangements. In bargaining, the focal time frame provided the impetus needed to finally reach an agreement, as recalled by Ban Ki-Moon: "No matter how much time is allotted, no matter how anodyne the

[63] Evans and Steven 2012, 4. [64] IISD 2012a, 21.
[65] For a discussion of this process, see: Kamau et al. 2018, 43–45.

subject, people will find it necessary to debate a point until it's time to come together or pull the plug. [Rio+20] was the moment."[66]

The Rio+20 action agenda, contained in the conference outcome document, *The Future We Want*, called on the UN General Assembly to take decisions on key conference outcomes.[67] These included upgrading UNEP's Governing Council to include universal membership, an increase in funding for UNEP, the effective replacement of the CSD with the HLPF, and the initiation of a process to specify and approve the Sustainable Development Goals. Other key outcomes included an improved science-policy interface, measures to operationalize a ten-year Framework of Programmes on Sustainable Consumption and Production, capacity building, and a consolidation of UN headquarters functions in Nairobi, Kenya. These measures would be adopted by the UN General Assembly on December 21, 2012.[68]

While UNCSD outcomes were widely panned as a disappointment at the time,[69] the institutional changes realized at the conference were more substantive than many acknowledged in 2012 and have aged well. Adjustments to UNEP and the creation of the HLPF have so far served as worthwhile and constructive improvements to UN environmental cooperation. The SDGs, moreover, have grown in importance over time and may serve as the most significant cooperative legacy of the second Rio earth summit. More importantly, as noticed by some observers at the time, Rio served as a vital platform for unprecedented social engagement that stretched beyond governments in the promotion of sustainable development, particularly encompassing the private sector.[70] This was, with the benefit of hindsight, a harbinger of things to come. A chief virtue of the SDGs and post-Paris climate governance has been a reach that extends well beyond government circles.[71] Such engagement is reflective of a wider diffusion of power within the international system. Furthermore, for those comparing Rio+20 with the international legal outcomes of UNCED, especially the Rio conventions, UNCSD did not measure up. The 2012 conference, however, was reflective of a broader shift underway since the early 1990s away

[66] Ban 2021, 255. [67] UNGAR 66/288 2012. [68] UNGAR 67/213 2012.
[69] See, for example: Halle 2012; Marx 2012; Pearce 2012.
[70] See, for example, Howard 2012.
[71] For a discussion, from an ethical perspective, of the Paris Agreement, see: Chan 2016.

from formal, legally binding institutional arrangements.[72] The approach, reflected Stanley Johnson, involved deploying "[m]ore carrot and less stick."[73]

Significantly, there were no clear changes in material or ideational/normative conditions in the months or years immediately preceding UNCSD that persuasively explain the timing of change in UN environmental institutions in 2012. Many environmental problems had been long-standing, often lasting decades. The incentives to realize change had accumulated progressively. The institutional limitations targeted in Rio had been long recognized by governments and, in respect of UNEP, the menu of institutional alternatives was well established. In addition, particularly concerning the proposed SDGs, new ideas made a difference in Rio. Rather than provoking coordination to institutionalize new norms and ideas, however, governance innovation was *provoked by* coordination. Convergent expectations in the context of a Temporal Focal Point stimulated Colombian thinking and entrepreneurship surrounding the Sustainable Development Goals. As the focal time frame neared, intensive political and analytical investments enabled intergovernmental action. As in 1992, the timing of institutional change is therefore explained by a rather arbitrary stimulant: a focal date around which actor expectations could converge. The "spirit of Rio," a manifestation of the coordinative power of an otherwise meaningless juncture on the temporal continuum helped to generate political momentum and a sense of opportunity to seize long available change incentives. The outcome was, in the reflections of Maria Ivanova, "subtle, yet significant."[74] Despite low expectations, Rio had produced a "definitive turning point in sustainable development politics."[75]

2012–2021: Realizing Sustainable Development

The post-Rio+20 phase of UN environmental politics saw the implementation of a significant transformation in patterns of cooperation. UNEP's Governing Council was transformed into the UN Environment Assembly (UNEA), featuring universal membership.

[72] Manulak 2021a, 2021b; Manulak and Snidal 2021; Vabulas and Snidal 2013.
[73] Johnson 2012, 252. [74] Ivanova 2013b, 211–231.
[75] Dodds and Nayar 2012, 2.

Measures were also adopted to strengthen its funding arrangements. The first UNEA was held in Nairobi from June 23 to 27, 2014, featuring high-level participation. At the time of writing, UNEA has had four sessions. The work of the Environment Assembly continues to be supported by a Committee of Permanent Representatives to UNEP.

The HLPF was established in July 2013 with the adoption of UN General Assembly resolution 67/290.[76] The HLPF was constituted under both the UN General Assembly and the UN's Economic and Social Council. The ECOSOC meetings of the HLPF, which include ministerial participation, are held on an annual basis. Meetings at the head of state or government level are held on a quadrennial basis. The HLPF aims to provide a central point of coordination for the implementation of sustainable development within the UN system and, to this point, has represented an upgrade over the CSD. With limited authority and modest resources, the forum relies on its legitimacy and political weight to "orchestrate" UN sustainable development activities.[77] To date, the HLPF has focused principally on reviewing progress on the follow-up and implementation of the 2030 Agenda for Sustainable Development.

Negotiations over the SDGs were more elaborate, moving swiftly to build off of the momentum generated by Rio+20. The OWG on SDGs completed its work in two phases. In the first, stretching from March 2013 to February 2014, it conducted extensive stocktaking, embarking on a detailed analysis of the current global context through meetings with experts and stakeholders. This phase of the OWG proceedings was vital to the effective depoliticization of the process, helping to change perceptions around key issues.[78] In the second phase, which ran from February to July 2014, the OWG deliberated on the proposed SDGs. While Rio+20 mandated a thirty-member OWG, more than seventy states participated via the formation of thirty small group "troikas" that allowed states to effectively "share" their seat on the OWG. This process, combined with the decision of the OWG's co-chairs, Macharia Kamau of Kenya and Csaba Korösi of Hungary, to hold onto the "pen" in negotiations, contributed to the adoption of the goals.[79] This innovative approach was seen as vital to developing an

[76] UNGAR 67/290 2013. [77] Abbott and Bernstein 2015.
[78] For a discussion, see: Kamau et al. 2018, 101–103.
[79] See: Kamau et al. 2018, 158–160.

effective, consensus-based process that resulted in successful negotiations. Given the fraught politics and deep concerns that had accompanied discussions over the SDG concept during the UNCSD preparatory process, the productive work of the OWG was notable.

Emerging from the OWG process were seventeen Sustainable Development Goals, ranging broadly across the social, economic, and environmental spheres. These include bringing an end to poverty in all its forms, ending hunger and improving food security, ensuring healthy lives, promoting education, achieving gender equity, improving access to clean water and sanitation, enabling access to energy, promoting inclusive and sustainable economic growth, building resilient infrastructure, reducing inequality, making cities safer, rendering consumption sustainable, tackling climate change, conserving oceans and marine resources, protecting and restoring terrestrial ecosystems, fostering peaceful and inclusive societies, and strengthening the means of implementation for the global partnership for sustainable development. The target date for achieving most of the SDGs is 2030. The 2030 Agenda for Sustainable Development, with the SDGs at its heart, was adopted unanimously at the United Nations Sustainable Development Summit in September 2015.[80] The summit, headlined by US President Barack Obama and Pope Francis, was accompanied by attractive and colorful marketing of the goals. To promote implementation, governments agreed to a series of more specific targets and indicators to measure progress in implementing the goals.[81]

The SDGs are far more comprehensive than their predecessor, the MDGs, and have the virtue of seeking to more fully leverage trillions of dollars available through the private sector. Indeed, they helped to advance a process of enhanced UN engagement with the private sector in the new millennium, including through public-private partnerships and the Global Compact. While environmental factors played a limited role in the MDGs, the SDGs more fully express the interconnectedness of development and earth systems. Social dimensions of development are also highlighted, including governance, conflict, and human rights. The goals call upon developing countries to mobilize their domestic resources, rather than focusing chiefly on development assistance from donor states. Significantly, the SDGs apply equally to all member states, helping to ensure their relevance in all countries.

[80] UNGAR 70/1 2015. [81] UNGAR 71/313 2017.

While widely lauded, implementation of the goals faces important challenges. In articulating 169 targets underpinning implementation of the SDGs, the priority-setting effort involved very little actual priority-setting. With so many goals and targets, moreover, tensions between different goals are inevitable. Some, for example, have argued that the SDGs provide an insufficient focus on the environmental pillar of sustainable development.[82] A review of SDG implementation ahead of the September 2019 HLPF meeting, occurring before the massive socio-economic disruptions caused by the 2020–2021 coronavirus pandemic, showed that no country was on pace to meet the SDGs and few will achieve even half of them.[83] If governments are to meet the SDGs, therefore, there is a need for significantly improved domestic implementation. A "decade of action" on the SDGs, which commenced in 2020, seeks to spur progress.

Even as they were approved by governments in 2015, there was strong recognition of the outsized role that addressing global climate change would have on implementing all of the SDGs. Accordingly, even then, the Paris Climate Change Conference loomed large over the proceedings. Thus, the 2015 United Nations Climate Change Conference and implementation of the resulting Paris Agreement, which seeks to limit global warming "well below" 2°C and "pursue efforts" to limit the temperature increase to 1.5°C, is essential for achieving the SDGs. Along with the approval of the SDGs and the publication of *Laudato Si'*, Pope Francis' influential encyclical on the environment, the Paris climate agreement capped off a dynamic year in global environmental politics.[84] As states approached "Paris+5," the fifth anniversary of the Paris conference, significant action was required to meet agreed-upon targets. Indeed, parties to the agreement have faced setbacks in implementing existing commitments and steps to enhance Nationally Determined Contribution have lacked ambition.[85] At time of writing, there are, however, reasons for optimism. Along with the commitment of the new US administration of President Joe Biden to rejoin the Paris climate agreement, pledges among several major emitters in 2020 to target "net zero," carbon neutrality, by 2050 offers hope that cooperation to mitigate climate change may be gaining momentum.

[82] Hickel 2020. [83] Our World in Data 2020. [84] Pope Francis 2015.
[85] American Geophysics Union 2019.

Efforts to strengthen UN environmental institutions also continue. Ongoing challenges facing coordination, particularly between MEAs and UNEP, continue to hinder governance efforts. Gaps have also been noted in the implementation of International Environmental Law. To tackle these challenges, the UN General Assembly created an open-ended working group in 2018 to provide advice on a possible "Global Pact for the Environment."[86] The working group reported on gaps in environmental law and environment-related instruments, focusing particularly on the need for strengthening national-level implementation of legal instruments and policies. These recommendations form the basis for negotiations underway at UNEP's Committee of Permanent Representatives toward a political declaration for the planned high-level "Stockholm+50" meeting, scheduled in 2022. Reportedly talks have concentrated on enhanced UN system-wide coordination and targeted measures to strengthen implementation of environmental law.[87]

Conclusion

States adopted substantial institutional changes at the 1992 Rio Earth Summit that sought to institutionalize sustainable development within the UN system. Combined with continued fragmentation of UN environmental institutions owing to the proliferation of MEAs, these changes created a series of institutional ambiguities that rendered cooperative structures suboptimal. The institutionalization of sustainable development within the UN system was thus incomplete. A series of efforts throughout the 1990s and early 2000s aimed to address these challenges, including the 1997 Nairobi declaration and the 2000 Malmö declaration. While problems were acknowledged and a series of institutional alternatives were in circulation, divergent temporal expectations led to the persistence of the institutional status quo. States were unable to adopt available change incentives.

As preparations began for the 2002 WSSD, which included a focus on institutional questions, a discrete time frame for coordinated action neared. A number of other important global talks, including the United Nations Conference on Financing for Development, however,

[86] UNGAR 72/277 2018.
[87] International Institute for Sustainable Development 2020a.

diminished the uniqueness of the Johannesburg summit. "Conference fatigue" set in. In addition, the September 11, 2001 terrorist attacks in the United States, occurring while the preparatory process for WSSD was underway, had important implications for the global political context. As attention shifted to the war in Afghanistan and to international counter-terrorism, this transformative event diminished the conspicuousness of WSSD. A temporal divergence of expectations resulted and international actors maintained a largely defensive posture in institutional bargaining over the future of UN environmental cooperation.

Divergent expectations continued through the first decade of the new millennium with institutional suboptimality persisting as preparations began for the 2012 UNCSD. While hopes were not high for Rio +20, momentum built in the final stages of the preparatory process. Combined with Brazilian efforts to enhance the visibility of the conference, the legacy of UNCED and the "spirit of Rio" led to a sense of exceptionality in talks. UNCSD became a Temporal Focal Point in UN environment politics. Governments moved decisively to universalize the membership of the UNEP Governing Council and adopt strengthened financial arrangements. The conspicuous time frame also provided a focus for intergovernmental initiative, reflecting substantial political and analytical investments in change processes. The underperforming CSD was folded and a new HLPF was created to provide a high-level vehicle for interagency and intergovernmental coordination to implement sustainable development. A Colombian initiative to articulate a set of post-2015 Sustainable Development Goals, which first emerged approximately a year before UNCSD, was also adopted at the conference. Negotiations provide ample evidence of temporal-pacing, illustrating the power of Temporal Focal Points in global affairs. The cases reviewed in the chapter, therefore, provide strong outcome-level support for the theoretical framework detailed in Chapter 2.

8 *Conclusion*

Social scientists have recently placed a good deal of emphasis on understanding the drivers of institutional change. Despite wide interest, important questions remain. One promising research agenda focuses on issues of timing and temporality. Time has been a neglected dimension in the institutionalist literature. When it has been addressed, questions of time have been evaluated chiefly from the perspective of sequence. Historical institutionalists have, for example, demonstrated persuasively that events that happen earlier in a causal sequence can have long-lasting impacts on cooperative relations, including through lock-in and increasing returns. This emphasis on sequence is complemented by the approach adopted in this book, which provides a novel conceptualization of temporality. The capacity of international actors to coordinate their activities and expectations at discrete points along the temporal continuum has a critical impact on institutional dynamics.

This book demonstrates the importance of temporal factors in institutional life. Even when shifting exogenous conditions create incentives to alter institutions, temporal coordination problems can hinder greatly the realization of Pareto-improving institutional change. In order for change to occur, a large number of moving parts must come into alignment at one point in time. Often, hundreds of states, subnational actors, nongovernmental groups, media, and expert bodies must engage in the reform process simultaneously. Coordination is extremely difficult in international fora. For each of these actors, backing change is a risky enterprise, involving the use of scarce analytical and political resources. If one side bargains cooperatively and others do not, the cooperative player wastes resources and can be exploited in future negotiations. It is, therefore, in the best interest of each actor to pursue change only when they believe that others will do so.

Crucial to overcoming coordination failure is the emergence of a temporal convergence of expectations – a specific time frame when actors expect others to expect to be expected to coordinate. It is this expectation that makes major political and analytical investments in bargaining a welfare-improving proposition. Convergent expectations can reshape fundamentally the information environment and the balance of political support for institutional reform. To a far greater extent than existing theory allows, this is a question of timing. Convergent expectations can be triggered by exogenous shocks and Temporal Focal Points. By easing assurance problems, a temporal convergence of expectations produces a spike in analytical activity, increased political entrepreneurship, and a shift in bargaining strategies.

In this concluding chapter, I summarize the findings of the preceding empirical chapters. The detailed case study investigations contained in this book, concentrating on six distinct phases in the history of United Nations environmental governance, support the plausibility of my framework. In four of the case studies, a temporal convergence of expectations increased the likelihood of change. Conversely, the absence of prominent temporal landmarks around which expectations could converge contributed to ongoing coordination failure and the persistence of extant institutional arrangements throughout much of the periods outside of these exceptional phases. The second part of this chapter examines the contributions of the theoretical framework to the study of institutional change and discusses policy implications. I conclude with some observations relevant to current global environmental politics.

Overview of Findings

The empirical research contained in this book covers more than fifty years of the history of United Nations environmental governance. Using newly released archival documents, an extensive program of interviews, and contemporary assessments, it provides a highly detailed analysis of the record of change in an important realm of institutional activity. In addition to making an original contribution to historical knowledge of the subject area, the case studies selected have *least-likely* qualities for a theory of temporal coordination. Coordination in an institutional setting with such a large and heterogeneous membership is extremely difficult. The near continuous presence of cooperation problems and the emergence of multiple socio-normative transition points, furthermore, make these *most-likely* cases for other theoretical traditions. Since my analysis covers the intervening years between my main cases, as well as instances where reform was pursued unsuccessfully, the case studies achieve variation in my dependent variable.

As world environmental problems mounted in the 1950s and early 1960s, states instituted a variety of measures on an ad hoc basis to cope with planetary issues through international cooperation. By the mid-1960s, however, continued technological and demographic change necessitated a more comprehensive approach. Within the United Nations, the United States took the lead in advocating an international response to environmental problems in 1963. Although joint action to swiftly tackle environmental problems was in the interest of all, states faced important risks in moving to a new institutional equilibrium. In addition to the investment of scarce resources in reform efforts, cooperative bargaining that was not reciprocated could have weakened the negotiating position of cooperative players in future talks. For example, industrialized countries feared environmental measures that, in effect, provided developing countries with a license to pollute and exploit their natural resources unsustainably. The rise of "pollution havens" in the global South could hurt the economic position of industrialized countries and negate the anticipated environmental benefits of cooperation. Developing countries feared that environmental measures could hinder their development and freeze the existing unequal distribution of global wealth. While these fears may have been exaggerated somewhat on both sides, all countries recognized that the stakes were high in North/South environmental talks. Coordination problems hampered efforts to pursue institutional change.

This situation changed with the emergence of a Temporal Focal Point: the 1972 United Nations Conference on the Human Environment (UNCHE). The "Stockholm conference," originally seen as a largely technical affair, quickly became a landmark political event. The belief among all actors that all other actors regarded UNCHE as highly important made the conference a focus for institutional initiative. UNCHE thus precipitated a temporal convergence of expectations. Actors expended heavy political and analytical resources in efforts to recast institutional arrangements in time for the 1972 gathering. The search for a suitable basis for reform gave great power to the findings of the 1971 Founex seminar on environment and development. While the Founex report did not break new conceptual ground, it carefully balanced the priorities of developed and developing countries. Most importantly, it arrived at a moment in time when states were seeking a suitable basis for reform. The Founex report shaped important elements of the Stockholm conference action plan and declaration. The institutional vision of Founex also manifested itself in the organizational measures that were adopted in Stockholm and eventually formalized in December 1972 at the United Nations General Assembly.

The new United Nations Environment Programme (UNEP) was designed to coordinate and catalyze environmental action within the UN system. The programme was not just the UNEP secretariat and Governing Council, but the sum of all the environmental activities of member states, nongovernmental and intergovernmental organizations acting in the UN context. This expansive vision encountered important difficulties in its first decade. To function effectively, UNEP required close cooperation among states, the UNEP secretariat, and UN agencies. Conflict among states and within the UN system contributed to a mixed record of action between 1972 and 1982. Voluntary contributions to UNEP's Environment Fund began to flag and the level of governmental representation at annual Governing Council sessions diminished. States focused obsessively on administrative and budgetary questions, neglecting vital programmatic issues. By the early 1980s, some actors began to call for reform.

With the approaching tenth anniversary of the Stockholm conference, UNEP's Executive Director, Dr. Mostafa Tolba, proposed a 1982 UNEP "Session of a Special Character" (UNSSC). The "Nairobi conference," Tolba hoped, would help regain the momentum and the coordinative impetus of UNCHE. UNEP and its enterprising executive director put great effort into making the Special Session a prominent political event. Tolba traveled the world, delivering speeches, giving interviews, and drawing attention to the second UN environment conference. UNEP published an extensive set of documents to influence world opinion at the time. Independent of UNEP's efforts, the tenth anniversary of the Stockholm meeting caught the eye of important actors. Many treated 1982 as an exceptional year in UN environmental politics. The Nairobi conference, therefore, served as a Temporal Focal Point, precipitating a convergence of expectations.

Numerous countries launched political initiatives to meet the focal time frame associated with the coming conference. The most prominent of these was an African group proposal to alter UNEP's mandate so that it could conduct its own operational activities. Despite the efforts of Kenya and Tunisia, this initiative was ultimately defeated. The proposal was unsuccessful because underlying conditions had not changed greatly since 1972. The list of environmental concerns facing states had not produced incentives sufficient to motivate change. Many environmental questions that were regarded as dire in 1972 were seen as less urgent in 1982 because of scientific and technological advances. The relative stability of external conditions meant that joint gains were not available to form a basis for a regime transformation in Nairobi.

The UNSSC case study provides strong evidence for the importance of factors exogenous to the institution itself in forming a substantive basis for institutional change. While UN environmental institutions faced many organizational limitations, these proved to be an insufficient driver for reform in 1982. Despite the failure of change, the Nairobi case supports the predictions of the model presented in this book. The prominent time frame stimulated significant analytical and political investments in the exploration of institutional alternatives. It was only *after* a spike in institutional diagnostics and analysis that it became clear to actors that a bargaining range did not exist.

This situation began to change just months after UNSSC, when the threat of a Mexican government default on its external public debt precipitated the Third World debt crisis. Many developing countries, particularly in Latin America and Africa, serviced their unmanageable debt through the unsustainable exploitation of their natural resources. Through the mid-1980s, there was a significant increase in the rate of environmental deterioration, including a worsening of desertification and deforestation. A string of environmental disasters, including the Bhopal industrial accident in 1984 and the Chernobyl nuclear meltdown in 1986, reinforced a sense of world environmental crisis. The shift in underlying conditions provided a basis for joint gains. It was in this context that the World Commission on Environment and Development (WCED) delivered its important report, *Our Common Future*, in 1987.

The "Brundtland report" sponsored an "alternative agenda" that amounted to sweeping institutional change. The commission advocated the complete integration of environmental considerations with the social and economic sectors of governmental and intergovernmental organizations. The existing "react and retrofit" approach to environmental management could not, the commission argued, address planetary problems. Instead, states would need to move to an "anticipate and prevent" approach, engaging central governmental agencies in the promotion of "sustainable development." At the international level, this entailed a greater role for international financial institutions, such as the World Bank and the regional development banks, in global environmental governance. Within the United Nations system, the UN secretary-general should serve as the main vehicle for interagency cooperation and coordination.

When the Brundtland report was delivered to the UN General Assembly in autumn 1987, many – including Brundtland herself – hoped that the reform process would begin immediately. Worsening environmental conditions – detailed in the commission's report – justified this expectation. Although states welcomed the report and its conclusions, they did not elect to engage with WCED's institutional recommendations in the UN context in 1987. The paradoxical approach of states to the commission recommendations can be explained by the existence of temporal coordination problems. There was a significant degree of ambiguity over the timeline for follow-up on the commission report. Some expected negotiations to begin in 1987, while others looked toward a series of regional conferences or even a notional international conference in 1992.

Although exogenous shifts had opened up joint gains, leading states had much to lose in institutional bargaining. Developing countries most feared the rise of green conditionality in the provision of development assistance. Developed countries, on the other hand, were reluctant to embrace the still vague concept of sustainable development, fearing that it could provide a justification for developing country exemptions from environmental efforts. Although each side had much to gain through institutional change, therefore, they risked compromising their future negotiating position if they bargained cooperatively. Rather than a joint search for viable means of institutionalizing the sustainable development concept within the UN system, defensive posturing won the day and change did not occur.

The WCED case study demonstrates the importance of convergent expectations in the realization of institutional change. Exogenous shifts had opened up joint gains that could be realized in 1987. However, divergent expectations contributed greatly to states' unwillingness to move definitively to institute new patterns of cooperation. Though governments widely embraced sustainable development at the national level and within other, smaller intergovernmental fora, they approached UNGA negotiations cautiously. North/South talks lacked direction and purposeful bargaining did not take place. The Brundtland case study also highlights the importance of gradual exogenous shifts in patterns of change. The progressive deterioration of environmental conditions in the 1980s did not provoke an institutional response for close to a decade.

The recommendations of the Brundtland Commission would be given a second life when the UN General Assembly scheduled the United Nations Conference on Environment and Development (UNCED) for June 1992. Despite urgent problems, temporal coordination dilemmas had hindered institutional change through the 1980s. The Rio conference quickly altered the situation. The Earth Summit became an event of unparalleled profile. No other environmental forum competed with the UN conference for the attention of actors. The twentieth anniversary of UNCHE, which the Rio conference marked, was a unique and highly conspicuous moment in UN environmental politics. UNCED, therefore, served as a Temporal Focal Point.

As the conference neared, there was an urgent search for solutions that could address the changed problem-setting. States behaved *as if* a major temporal discontinuity would follow the Earth Summit. No serious attention was given to the possibility of continuing negotiations after UNCED, for example. Although many institutional options were initially entertained, the WCED recommendation provided an almost irresistible focus for talks. Bargaining on institutional questions centered on altering the decision-making processes of the World Bank-operated Global Environment Facility and on creating a UN Commission on Sustainable Development (CSD). The authority of the UN secretary-general was also enhanced to improve interagency coordination. Sustainable development, a concept popularized by *Our Common Future*, served as the conceptual basis for agreement on institutional questions in Rio de Janeiro.

Without the Temporal Focal Point framework, the timing of institutional change in 1992 is puzzling. There was no exogenous shock or parametric shift in close temporal proximity to UNCED that persuasively explains the timing of change.[1] While underlying conditions had shifted significantly with the arrival of the Third World debt crisis in August 1982, states were unable to coordinate purposeful negotiations to realize joint gains. For sociological institutionalist approaches, the fact that it took states almost five years to act on key WCED recommendations, which were by the early 1990s already widely implemented in some other fora, is perplexing. The best explanation for the timing of institutional change in 1992 was that it coincided with the twentieth anniversary of the Stockholm conference. Relevant actors used the prominent anniversary to coordinate negotiations to overcome the risk-dominance of the prevailing equilibrium and achieve Pareto-improving institutional change.

In some respects, UN environmental cooperation was a victim of its own success at Rio. A series of institutional ambiguities, particularly concerning the role of UNEP and the CSD, surfaced following UNCED and hampered cooperation. The continued proliferation of multilateral environmental agreements added to fragmentation and coherence problems in UN environmental cooperation. As a result, states considered several proposals for institutional change in the late 1990s and early 2000s that aimed to complete the institutionalization of sustainable development within the UN system, including upgrading UNEP to the status of a UN specialized agency. As the new millennium approached, preparations began for a summit-level conference to mark the tenth anniversary of UNCED. The World Summit on Sustainable Development (WSSD), held in August/September 2002 in Johannesburg, South Africa, presented an opportunity for coordination.

The Johannesburg summit targeted a review of the implementation of Agenda 21, the Rio action plan, including a focus on institutional questions. While WSSD was a discrete moment in time, it lacked the uniqueness or conspicuousness of the other big UN conferences analyzed in this book. Its uniqueness was diminished by a series of

[1] As noted in Chapter 6, the end of the Cold War did not have a major impact on cooperative potentialities in the environment sphere. The major fault lines on global environmental issues had always been North/South, not East/West.

international conferences held in the period, leading to "conference fatigue" among international actors. The terrorist attacks on the United States on September 11, 2001, occurring while conference preparatory processes should have been hitting their stride, consumed many international actors and limited the conspicuousness of the South African meeting. WSSD does not, therefore, meet the definition of a Temporal Focal Point. Action was not taken to address the institutional limitations previously acknowledged by governments. The Johannesburg summit did not result in significant institutional change.

Despite the presence of incentives to alter UN environmental institutions, a temporal divergence of expectations stretched into the early 2000s. Ongoing efforts to precipitate coordination brought disappointing results, as states continued to assume a defensive stance in intergovernmental treatments of the significant challenges facing UN environmental institutions. The institutional status quo was persistent, as levels of government participation in UNEP and the CSD began to diminish. In the context of a crisis in the Eurozone and a US election, expectations for the United Nations Conference on Sustainable Development (UNCSD), to be held in Rio de Janeiro, in June 2012, were low.

As the "Rio+20" conference neared, however, it became an increasingly conspicuous phase in UN environmental affairs. International actors regarded the Brazilian meeting as the best possible vehicle for addressing long-standing institutional suboptimality and dysfunction, realizing available change incentives. Combined with a concerted effort by the Brazilian government to increase the visibility of the meeting, the "spirit of Rio" helped to generate a sense of exceptionality leading up to the conference. UNCSD thus served as a Temporal Focal Point in the history of UN environmental governance. Governments acted to adopt institutional alternatives that had long been in circulation, such as approving universal participation in the UNEP Governing Council and enhanced funding arrangements for the programme. The conference served as a platform also for significant institutional initiative, including the replacement of the CSD with a High-level Political Forum (HLPF) on Sustainable Development. A Colombian initiative led to agreement on a process to articulate a series of Sustainable Development Goals (SDGs) for the post-2015 period.

One of the important commonalities between the UNCHE, UNCED, and UNCSD case studies is the relevance of coordination along two dimensions: bargaining space and time. The coincidence of the Stockholm Temporal Focal Point and the Founex seminar's conclusions were crucial to realizing institutional change in 1972. While the Founex report brilliantly met the central negotiating challenge of the Stockholm preparatory process, its temporal proximity to the fast-approaching conference added greatly to its power. In a similar way, the institutional impact of the Brundtland Commission in the UN context ultimately owed much to UNCED. The urgent search for solutions to the bargaining dilemmas of the Earth Summit facilitated the institutionalization of sustainable development in the UN system. UNCSD became a focus for international actors in devoting significant attention to long-available institutional alternatives and pursuing new cooperative concepts, such as the HLPF and the SDGs. Key to institutional change in these cases, therefore, was the ability of states to coordinate their expectations along two dimensions. The temporal dimension was imperative to this process.

Taken together, the cases provide strong evidence to support the model of coordination elaborated in Chapter 2. The role of exogenous shifts in giving rise to the incentives that ultimately fueled institutional change is apparent in the case studies. It was actors' shared interest in halting planetary decline that provided the substantive basis for alterations to institutional rules and procedures in 1972, 1992, and 2012. In the case of the 1982 Nairobi conference, the absence of significant incentives for change explains reformers' inability to transform UNEP's mandate. While there were clear organizational limitations, there was no substantive basis for agreement. Though imperfect, relevant actors regarded existing institutional measures as the best that could be achieved in light of prevailing international conditions.

The model's emphasis on the role of convergent expectations in time on patterns of institutional continuity and change is also supported by the evidence. The importance of temporal landmarks is seen clearly in all six cases. In the Stockholm and the two Rio case studies, states had been slow to adopt measures to address important shifts in underlying conditions. Temporal coordination problems had hampered reform efforts significantly contributing to persistent institutional suboptimality. The emergence of a TFP changed the situation. Coordination problems were eased by the presence of a focal point, enabling a precipitous increase in political and analytical activity. Actors bargained purposefully, seeking to realize change within the focal time frame. While there was no substantive basis for agreement, Temporal Focal Points had a similar impact in 1982. In the Nairobi and both Rio case studies, the timing of TFPs had little to do with exogenous or ideational developments. Institutional change was enabled by the arrival of a prominent anniversary that served as a basis for temporal coordination. In 1987 and 2002, divergent expectations slowed the realization of institutional change, despite the presence of incentives to alter institutional arrangements. It took a prominent temporal landmark, the twentieth anniversaries of UNCHE and UNCED, for actors to negotiate purposefully and increase significantly their political and analytical investments in change processes.[2]

Relationship with Existing Literature

Before proceeding to a discussion of policy implications, a brief analysis of how the proposed approach fits into the broader theoretical literature on institutional change is in order (as summarized in Table 8.2). Since international relations theory lacks a full conceptualization of temporal factors, it is unsurprising that leading theoretical treatments of institutional change leave the question of timing largely implicit.[3] These strands of institutional theory stand to benefit from an increased focus on the implications of temporal factors and risk-dominance for the coordination of institutional change. In pure rational choice accounts, institutions change tracks efficiency. Delay is irrational, akin to opting for Pareto-inferior institutional arrangements. Even when one takes into account the role of transaction costs in preserving institutional stability, thus assuming that exogenous transformations must be of a somewhat greater magnitude than when one assumes pure rationality, the timing of change is still driven principally by changes in external parameters.[4] Institutionalists who emphasize the role of transaction costs in preserving institutional

[2] Process-based conclusions for the latter two cases do not have the benefit of the more detailed archival assessments of the first four case studies. As a result, conclusions in this book have focused chiefly on an outcome-based, rather than process-based, assessment. Preliminary process-based analysis is, however, included in Table 8.1 for completeness.

[3] For a detailed analysis of the small, but growing literature on temporal factors in international institutions, see: Jupille and Caporaso 2021. See also: Fioretos 2017; Hom 2020; Rixen et al. 2016.

[4] Keohane 1984.

Table 8.1 *Summary of case studies*

Period	Outcome-based Implications	Process-based Implications
Pre-1972	• Incentives present to alter institutions, at least from 1963 to 1971, but no change. • Convergent temporal expectations around 1972 Stockholm conference triggered institutional change.	• Significant increase in analytical investments focused on the state of the world environment. • Creation of working groups concentrated on institutional questions and the devising of institutional alternatives. • Evidence of temporal-pacing, with bargaining predicated heavily on the Founex solution to Stockholm negotiating dilemmas.
1973–1982	• UNEP faced organizational limitations, but environmental conditions had not changed greatly since 1972. As a result, few incentives existed to alter institutional arrangements. • 1982 UNSSC facilitated coordination, but lack of incentives to motivate change led to institutional continuity.	• Analytical investments increased in the context of the Nairobi conference. • Political actors, including leading African states, made substantial political investments in reform efforts. Canada promoted the idea of an independent commission and Japan sponsored the Canadian initiative at UNSSC. • Bureaucratic jostling between US EPA and State Department influences US policy debates.
1983–1987	• Third world debt crisis creates economic pressure, significantly worsening the state of the world environment and creating incentives to alter institutions. • A temporal divergence of expectations persists, however, including at the UN General Assembly following the release of the Brundtland report.	• Divergent expectations led to posturing in institutional bargaining, with states reluctant to commit to the sustainable development formula.

1987–1992	• Exogenous conditions shift throughout 1980s, creating incentives to alter institutions. Divergent expectations lasted until momentum emerged toward the 1992 Rio "Earth Summit." • Convergent expectations build around Rio TFP, facilitating large-scale institutional change.	• Significant increase in institutional analysis in the wake of convergent expectations. • Major institutional initiatives advanced, including those related to the Commission on Sustainable Development. New bureaucratic units established. • Informal workshops organized at Aspen and Arden House to explore institutional questions. • Bargaining dynamics shift as states look to capture available incentives.
1993–2002	• Institutional ambiguities and the environmental effects of globalization presented incentives to adjust UN environmental institutions. • Divergent expectations prevented change.	• Divergent expectations appear to have led to posturing in institutional bargaining, with states reluctant to commit to address ambiguities and limitations in UN environmental institutions.
2003–2012	• Long-standing institutional suboptimality and fragmentation presented institutional change incentives. • The emergence of a Temporal Focal Point triggered a convergence of expectations around UNCSD, leading to institutional change.	• A temporal convergence of expectations led governments to swiftly adopt long available institutional alternatives. • A TFP led to political initiative, most prominently by the Colombian government, acting within the focal time frame. • Evidence of temporal-pacing appears, as states act swiftly to adopt agreement concepts that did not have the benefit of sustained intergovernmental consideration.

Table 8.2 *Theoretical comparison summary*

Approach	Timing of Institutional Change
Historical institutionalism	Following the arrival of critical junctures, or following more gradual patterns of endogenous change
Rational choice institutionalism	Following parametric change, taking into account a stability-preserving transaction cost cushion
Sociological institutionalism	Preceding or following the buildup of a critical mass of support, or following a reassessment of actor identities, interests, or cognitive scripts
Temporal Coordination Model	Following a temporal convergence of expectations

stability do not take into account the possibility that actors' willingness to assume transaction costs varies over time. If, however, one supplements rational choice insights with the approach proposed in this book, a set of predictions can be made about how the prospect of convergent temporal expectations might affect states' willingness to incur heightened transaction costs.

Historical institutionalists have noted the gaps associated with a purely rational choice approach to explaining change, assessing the implications of path dependence and "lock-in."[5] Giovanni Capoccia and R. Daniel Kelemen note that long phases of path-dependent institutional stability are "punctuated occasionally by brief phases of institutional flux referred to as critical junctures – during which more dramatic change is possible."[6] Ruth Berins Collier and David Collier define a critical juncture as "a period of significant change . . . which is hypothesized to produce distinct legacies."[7] Critical junctures open a rare window to the efforts of powerful and able actors to redesign or adjust institutions.[8] More recently, Capoccia has analyzed the impact of constructed cultural categories and the concentration of control over the timing of reform efforts on the likelihood that an institution will

[5] Fioretos 2011, 2017; Ikenberry 2000; Pierson 2004. Though rationalist theorists have regarded these as evidence also of the independent causal effect of institutions, see: Keohane and Martin 1995.

[6] Capoccia and Kelemen 2007, 341.

[7] Collier and Collier 1991, 29.

[8] Capoccia and Kelemen 2007, 360–361.

resist bottom-up endogenous change.[9] Hillel David Soifer has parsed the logic of critical junctures to identify permissive and productive conditions for change.[10]

An emphasis on temporal coordination dilemmas and transaction costs complements research on critical junctures, helping to explain institutional lock-in and change. The emergence of convergent expectations can account for *why* and *how* certain junctures become critical over the life of an institution, bringing together key insights from the rational choice and historical institutionalist traditions. Consequently, convergent expectations, including those stimulated by Temporal Focal Points, are an important cause of critical junctures. The approach proposed in this book entails the detailed, empirically rich analysis called for by historical institutionalists, effectively situating "politics in time." It shares with historical institutionalism a deep interest in how structure and agency interact to produce continuity and change in institutions. Its emphasis on how micro-motives influence macro-behavior, builds on agent-centric variants of historical institutionalism. Rational choice institutionalism also stands to benefit. By analyzing how states' willingness to assume transaction costs varies over time, my framework provides a plausible explanation for institutional lock-in and stickiness.

Accounts of a sociological bent also benefit from an analysis of the timing of institutional change. In theories of norm dynamics, for instance, change immediately precedes or follows cascades brought about by growing support for emergent norms.[11] The timing of institutional change is determined, therefore, by the success of norm entrepreneurs in persuading relevant players to support their normative enterprise. Few predictions are made concerning when the demand for such norms might be highest. Where institutional change follows a reassessment of actor identities, interests, or cognitive frames, insight is provided into when this reassessment might occur.[12] A focus on timing can sharpen the predictions of sociological accounts. Rather than norms and new ideas ebbing and flowing along their own timelines, the development of new norms and ideas may be stimulated by convergent expectations. If norm entrepreneurs time their efforts in response to certain temporal stimuli, then a fuller analysis of the relationship between entrepreneurship and the demand for such

[9] Capoccia 2016. [10] Soifer 2012. [11] Finnemore and Sikkink 1998.
[12] March and Olsen 1989; Wendt 1992, 391–425.

entrepreneurship would seem to be in order. If states are more receptive to new norms or more prone to adopting alternative cognitive frames at certain junctures, than this phenomenon would have important implications and should be explored. A potential market for new norms and ideas provided by the increased likelihood of successful coordination might, therefore, make the adoption and institutionalization of emergent norms more likely.

While this book focuses on large, discontinuous change, it also has implications for more gradual forms of change.[13] Coordination problems are present when actors contemplate less transformative change, particularly in large multilateral contexts. Indeed, when the potential gains from reform are more modest, as is often the case when change is incremental, the attraction of the stag-stag scenario is less pronounced, providing less incentive to incur heightened transaction costs. This is the case even if the transaction costs and disruptions associated with incremental changes are smaller. This can result in a widened gap between changes in the international environment and institutional arrangements.

Contribution of the Framework

The theoretical framework presented in this book complements and extends the literature on institutional change. By drawing attention to the severe coordination problems facing states in the realization of change, the model provides a novel explanation for the status quo bias of international institutions. Although coordination problems have received relatively little attention in the literature on institutional change, my evaluation demonstrates that they regularly slow regime transformations.[14] This focus can also provide insight on where and when significant change is most likely to arise. The empirical investigation contained in Chapters 3–7, for example, suggests that cooperative contexts characterized by high complexity face significant

[13] On graduate change, see: Mahoney and Thelen 2010; Young 2010.

[14] For some exceptions, see: Calvert 1998, 75–78. Calvert highlights the coordination dilemmas faced by the two houses of the United States' Congress. Coordination problems are faced in the context of multiple equilibria and the decision whether to confer and to assume the costs/risks associated therewith. Avner Grief and David D. Laitin (2004, 638) also discuss coordination issues briefly.

difficulties in coordination. Conversely, smaller, more homogenous institutions might be expected to change more rapidly when exogenous shifts create incentives for reform. In responding to the recommendations of the Brundtland Commission, for instance, smaller institutions, such as OCED, were able to respond more rapidly.

Convergent expectations explain the mechanism through which shocks precipitate regime transformations. In part because exogenous shocks are so commonly associated with large-scale change, it is easy to overlook the extent to which major alterations to global conditions provide a coordinative impetus for reform efforts. By theorizing the impact of coordination problems on change processes, it is possible to more fully analyze the impact of shocks on institutional equilibria. While exogenous shocks are a useful analytical concept, the high correlation between crises and change does not, moreover, necessarily mean that crises that precede behavioral modifications *cause* the change observed. Often, visible international crises can serve as conspicuous temporal signposts that enable coordination among actors. This is because the ingredients necessary for change accumulate progressively. The evidence presented suggests that Pareto-improving agreement opportunities are probably more frequently opened up by gradual exogenous shifts than by sudden shocks. International crises, which are assumed to be the cause of change, may simply bring into focus incremental shifts in key cooperative parameters that occurred in the past. Actors come to recognize that existing institutional arrangements are not an equilibrium and coordinate their expectations to realize joint gains. The scale of change may also exceed that expected through a strict application of the logic of shocks. This is also suggestive of the role of coordination dilemmas in institutional change processes.

A focus on the crystallization of temporal coordination creates greater room for the analysis of agency. Major sociological institutionalist, rational choice, and historical institutionalist contributions can assume a certain automaticity in change once prevailing ideational or material structures change. While existing institutional arrangements constrain behavior, there is room for creativity and strategic choice in the timing of reform efforts. The case studies examined in this book, for instance, underscore the entrepreneurial efforts of the UNCHE, UNSSC, and UNCED organizers in seeking to increase the visibility of certain moments in time. The findings suggest that institutional entrepreneurs do not just advocate particular agreement possibilities, but may also seek to provide a temporal focus for bargaining. Agenda *timing* is a central dimension of agenda-*setting*.

While Temporal Focal Points are generated exogenously, the pivotal role of entrepreneurs, such as Maurice Strong, is highly evident in the empirical chapters. By adroitly enhancing the conspicuousness of temporal foci, these actors did much to set in motion the emergent processes that helped to generate large-scale institutional change in 1972, 1992, and 2012. Although change was motivated by available incentives and facilitated by emergent dynamics, the role of such entrepreneurs was far from inevitable. Without their efforts, convergent expectations may not have resulted and the institutional status quo could have otherwise persisted for years (even decades). No focal point is inevitable. Indeed, my analysis of exogenous shifts, and the temporal coordination dilemmas associated therewith, suggests that even in the face of considerable change incentives, the status quo can endure for long stretches before coordinated action is achieved.

This conclusion highlights an endogenous aspect to Temporal Focal Points. As an ideal typical phenomenon, focal points always contain a significant exogenous element, while still allowing for agents to play a role in their crystallization. In practice, entrepreneurs can be real difference-makers in realizing coordination and are likely to be highly visible institutional players. From a policy perspective, crystallizers can seem to be the decisive actors on the institutional scene. These actors are able to recognize the arc of international history and perhaps help to accelerate its progression in the direction of institutional transformation. While true, it remains important not to lose sight of the impact of other factors in their success, such as preexisting change incentives and the relational social dynamics that allow expectations to converge at certain points in time.

A final contribution to the literature is a detailed and extended presentation of how institutions change over time. In contrast with the majority of existing accounts, which contain numerous diverse case studies, this book tracks more than fifty years of history within one institutional setting in detail. This approach allows for the analysis of issues of temporal coordination that play out over time. It is through this longitudinal study, identifying six sub-cases that allows for tighter control in analyzing variables of interest, that the role of convergent expectations and Temporal Focal Points come into view. This longer-term vantage point also casts new light on widely used concepts, such as exogenous shocks and critical junctures. A further benefit of such a detailed study is that it is better positioned to make a significant contribution to empirical and historical knowledge.

A longer-term perspective sheds new light also on the role of transaction costs in international cooperation. International relations scholars have focused heavily on the role of institutions in reducing transaction costs and on the implications of such costs for institutional design choices. This book has shown that transaction costs are also a central factor in explaining patterns of coordination, institutional continuity and change. States' willingness to assume heightened transaction costs varies over time and Temporal Focal Points serve as a crucial mechanism for economizing on these costs. An unwillingness to accept transaction costs – in the form of political and analytical investments in change processes – serves as a principal explanation for the status quo bias in many global institutions.

Policy Implications

A number of important implications flow from the preceding analysis. First, an emphasis on multi-actor coordination problems implies a more encompassing view of the set of important players in institutions than is typically adopted. In contrast with many existing accounts, which tend to focus on the role of the United States and a small number of other state-based actors in sustaining and altering cooperative equilibria, my model widens the analytical lens to explain how the loosely coupled activities of a larger subset of players can influence institutional change processes. The empirical evidence outlined in the book supports this concentration. In addition to smaller states, which can be highly consequential players, institutional change efforts implicate numerous agencies and departments *within* individual states. This is the antithesis of the hegemonic approach.

For policymakers and analysts, recognizing a wider circle of relevant actors is not just an analytical imperative. It is a practical one. When the same old players dominate institutional change efforts, you often get the same old outcomes. Even in the face of significant incentives to alter institutions, there is a strong status quo bias in institutional life. Large-scale institutional change thus necessitates a widening of the circle of actors involved to include agencies, departments, and levels of government that are not active in day-to-day national representation within a given institution. It is therefore not enough to involve senior decision-makers, engaging *vertically*. Connecting *horizontally* is also key. Much of the dynamism seen in the Stockholm, Nairobi, and Rio cases arose from the involvement of a broader subset of national actors in institutional processes, leading to a clarification of preferences.

Second, bargaining to alter institutional arrangements implicates a large number of non-state actors and their contributions can create avenues for large-scale change. In a parsimonious manner, the model presented in this book examines the impact of think tanks, academics, and nongovernmental organizations, alongside traditional state-based players, on the coordination of change. Whether it was the contributions of development economists via the Founex seminar or the insights gained through WCED and its public hearings, broad societal engagement enriched and impelled global talks. Such an approach holds special interest in a period when global power is becoming more diffuse. This conclusion is consistent also with experience in the lead-up to the 2015 Paris Climate Change conference. The covert "Groundswell Initiative," led by Tom Rivett-Carnac and quietly enlisted by UNFCCC Executive Secretary, Christiana Figueres, sought deliberately to build a multi-sector network of actors to make the Paris agreement more likely and more ambitious.[15] In what Joseph S. Nye terms, "the other global power shift" of our times, a broad spectrum of actors have moved to the international fore, transforming world politics. My research suggests that a more open public posture is vital to institutional modernization and renewal. The more divorced institutions are from public dialogue and debate, the more stale they become. When change is on the table, and – arguably – even when it is not, international institutions need to be brought into the public sphere.

For policymakers, such a recognition has clear implications for national and international resources devoted to public diplomacy and engagement vis-à-vis institutional questions. Upon taking over the UNCHE secretariat, Maurice Strong invested heavily in public outreach efforts. This included a marked willingness to use non-state actors as information and political force multipliers in bargaining. In part because of these efforts to crystallize the Stockholm Temporal Focal Point, the conference became a platform for the insights and

[15] Figueres and Rivett-Carnac 2020, xi.

activities of many, diverse governmental and nongovernmental players. This was fundamental to the conference's success. In many cooperative contexts today, a more open posture that is highly transparent and permeable to relevant stakeholders is essential to producing the kinds of political and analytical investments necessary to renew global institutions.[16] Rather than being seen as an add-on, public engagement is core business for international organizations and needs to be treated as such. For an example of effective recent outreach in the UN context, the innovative collaboration with filmmaker Richard Curtis in launching and marketing the Sustainable Development Goals helped to support broad-based public outreach. This effort involved slick marketing, evidenced by the colorful icons for the SDGs and celebrity engagements.[17] In many global institutions, however, there has been a public engagement deficit, making these bodies susceptible to governments that take credit for cooperative successes and blame them for unpopular measures or visible failures.

Third, while a wide array of actors should be considered, it is the broader pattern of relationships, rather than the attributes of those individual actors, that matters most in realizing change. Analyses of international affairs tend to concentrate on questions of polarity, emphasizing what divides states, rather than what brings them together. While such an approach has appeal, particularly in a period of increasing great power competition, it risks missing the forest for a few trees. My model suggests that the capacity of many, diverse players to concert their expectations is a necessary condition for the realization of change. In this, my assessment shares commonalities with Social Network Analysis in international relations.[18] By instead emphasizing the *connective tissue* of institutional life, the actors and shared perceptions that link sides, we can better explain sudden shifts in collective behavior that are often associated with critical junctures. In this sense, collective movements to alter institutions are an emergent property, irreducible to the activities of a small number of individual players. The analysis of such properties requires a shift from a concentration on individual states to a focus on the wider whole.

[16] For a related discussion, see: Slaughter 2017, 222.
[17] For a discussion, see: Kamau et al. 2018, 244–245.
[18] For two recent examples, see: Avant and Westerwinter 2016; Kahler 2009.

The case studies demonstrate the value-added of this analytical approach. As an international development administrator from a developed country with extensive ties to developing world economists and politicians, conference secretary-general Maurice Strong was positioned to enjoy significant influence in the crystallization of the UNCHE and UNCED Temporal Focal Points. Strong used his links to prominent development thinkers to orchestrate the Founex seminar and then promote its conclusions as a solution to the central bargaining dilemma facing states in Stockholm. Strong served on the Brundtland Commission and was well-positioned to advance the sustainable development agenda in the run-up to the Rio Earth Summit. Strong's rolodex included major non-state and sub-state actors, allowing him to act as a network "switcher" or "multiplex" node in the conference processes.[19] Rather than the possession of conventional power resources, therefore, pivotal dynamics played out as a consequence of broader relational patterns.

For policymakers and diplomats well-versed in traditional modes of interstate interaction, this perspective is challenging. Those that wish to precipitate change to institutional norms and rules, for example, should build a broad array of linkages, state-based and non-state, investing heavily in *bridging* capital.[20] Rather than directing or controlling change processes, as is often implied by an emphasis on great powers or hegemonic sub-groups, influential actors tend to orchestrate or mobilize. Within international organizations, efforts to reinforce the role of institutional connectors or bridging forces are paramount. The deliberate cultivation of multidimensional "task networks" is one possible means of strengthening connection.[21] Similarly, mechanisms that foster a "team-of-teams" approach to building productive linkage across various national divides can reinforce the modes of connection that favor institutional adaptation and change.[22] Finally, when electing states for governing bodies or selecting/appointing officials for international roles, a greater focus should be given to their functional position within actor networks and their capacity to facilitate coordination. While such an approach could diminish regional

19 For a discussion of multiplex nodes, see: Avant and Westerwinter 2016. On network "switchers," see: Castells 2012, 8–9.
20 On bridging versus bonding capital, see: Putnam 2000.
21 Slaughter 2017, 117–119. 22 McChrystal et al. 2015.

representativeness somewhat, the benefits of an increased emphasis on functional considerations make it worthwhile.

Temporal Focal Points and Global Environmental Politics Moving Forward

At the time of writing, the world continues to grapple with the effects of the coronavirus pandemic. Along with triggering a global health and economic crisis, the pandemic has illuminated the critical environmental challenges facing the international community, forcing global actors to come to grips with a new class of cooperation problems consistent with what is increasingly referred to as the Anthropocene. This new geological epoch is characterized by dramatic human impacts on earth systems.[23] The pandemic itself, widely believed to be sparked by zoonotic transmission, has brought into focus humanity's dysfunctional relationship with nature. Owing to the degradation of ecosystems, increased human encroachment on wild spaces, and exploitative food production practices, the risk of transmission of animal-borne viruses is growing rapidly. The emergence of pandemics is thus facilitated by many of the same human activities that have accelerated climate change and the loss of biological diversity.[24] Climate change has, moreover, been linked to the establishment and geographical expansion of zoonotic diseases.[25]

Relatedly, increased incidence of extreme weather, including drought and unprecedented heat waves, have demonstrated the implications of global climate change for human settlements. In its Sixth Assessment Report, released in August 2021, the Intergovernmental Panel on Climate Change (IPCC) was unequivocal in stating that the world is now observing the impacts of climate change brought about by human activities. Humanity has contributed already to 1.1 °C of warming since 1850–1900 and the report notes that warming is occurring more rapidly than previously thought. Without swift, immediate action to reduce emissions, it will be difficult for the international community to halt greenhouse gas emissions beyond 1.5 or even 2 °C. The report notes also that, at 1.5 °C warming, the world will experience increasing heat waves, longer warm seasons, and shorter cold seasons. If 2 °C warming is reached, extreme heat could more

[23] Bierman 2021. [24] IPBES 2020. [25] Germain et al. 2019.

often exceed critical tolerance thresholds for agriculture and health.[26] The social, political, and economic ramification of these levels of warming would be severe.

In his December 2020 "state of the planet" address, UN Secretary-General António Guterres summed up current planetary conditions: "the state of the planet is broken." Humanity, according to Guterres, is waging a "war on nature," causing natural systems to "strike back." This is not, Guterres maintains, just an environmental imperative. Planetary problems imperil, in a direct way, world development prospects and economic well-being.[27] UNEP, for its part, responded to the secretary-general's call, releasing in February 2021 its provocative report "Making Peace with Nature." The report pointed to the escalatory and self-reinforcing character of three global emergencies: climate, biodiversity, and pollution. The ramifications of these crises, noted the report, were significant for human health, as well as social and economic systems.[28]

Though UNEP's report treads lightly on organizational implications, essential to tackling these – and other – environmental emergencies is the need to respond institutionally in a more coherent and integrated fashion than ever before. The problems of the Anthropocene cannot be tackled in isolation. Research by Frank Biermann and colleagues shows that the environment sphere is more fragmented than many other areas of institutional activity.[29] Indeed, as UNEP's Executive Director, Inger Andersen, noted in a May 2021 interview, the system of multiple, largely autonomous governance arrangements for each environmental convention is more a product of historical legacies than of efficient design. If these structures were to be redesigned *de novo*, she conceded, it is uncertain whether such a setup would be replicated.[30] This fragmentary order may, furthermore, have already entered a phase of change resistant path dependence.[31] Current modes of environmental governance are moreover inefficient, risking the reproduction, at the national level, of the fragmentation seen at the global level, and imposing a burdensome reporting process on governments that results ultimately in a less encompassing perspective than is needed. This system gives rise to conflicts in implementation of

[26] IPCC 2021. [27] Guterres 2020. [28] UNEP 2021.
[29] Biermann et al. 2009; Biermann and Kim 2020. [30] Andersen 2021.
[31] Dryzek 2016.

intergovernmental agreements, sometimes leading institutions to work at cross-purposes or to compete for scarce funding.[32] As future environmental conventions are established, it will be necessary to prevent still further fragmentation. Even with the creation of the UN Environment Assembly following the Rio+20 summit in 2012, a step taken in large part to enhance system-wide coordination, fragmentation remains a major challenge and runs counter to the integrative thrust of UNEP's report and to the requirements of environmental governance in the Anthropocene. Indeed, it is the need for a more integrated approach that has helped to motivate the movement toward "nature-based solutions" to many current environmental challenges.[33]

Problems of coherence in the UN environment sphere are the tip of the iceberg. The degree of fragmentation within the broader UN sustainable development agenda is more profound still. The necessity of bringing social and economic systems into line with environmental imperatives has been demonstrated by the pandemic. Yet these very systems of integration, including the Sustainable Development Goals, have been eroded by the every-country-for-itself mentality that has at times defined the global response to the pandemic. Key indicators of progress on SDGs related to biodiversity and climate change, for instance, have shown a marked deterioration since early 2020.[34] Progress is hindered also by gaps in the diversity and use of environmental data and statistics that inform governmental policy responses. Outside of these environmental aspects of the global goals, overall progress on implementing the SDGs has halted and indeed reversed.

While these problems are not new and have increased in severity over time, their consequences have been demonstrated in dramatic fashion and exacerbated by the pandemic. The pandemic has fueled a "grow at all costs" mentality in some countries, leading to the neglect of social and environmental considerations.[35] At the same time, pandemic spending has contributed to a tightening of resources available for international initiatives, exacerbating competition for resources among international environmental organizations. Current systems of UN environmental governance are not well-suited to addressing the set

[32] On conflicts across regimes, see: Zelli 2011. In the area of climate governance, Robert O. Keohane and David G. Victor (2011) describe the loosely couple set of specific regimes as a "regime complex." It has so far proven highly difficult to build a more comprehensive regime in climate governance.

[33] See: IUCN 2021; UNEP 2019.

[34] IISD 2021.

[35] Harvey 2021.

of interconnected cooperation problems facing states today. An argument can – and indeed should – be made that the international community is in need of a "constitutional moment."[36] Numerous proposals have been in circulation for years, including converting UNEP into a specialized agency and transforming the UN's Trusteeship Council into a vehicle for governing the global commons and realizing sustainable development across the UN system. Some of these proposals would require substantial institutional change, including revisions to the UN Charter, but are needed to meet the set of cooperation challenges facing governments today. Incentives therefore exist to alter institutional arrangements. Whether this can be achieved through the UN, or via some other institutional venue, remains to be seen.

It is not the presence of incentives that is the chief impediment to change, however. The core issue is whether states can concert their expectations temporally. Can the multitude of actors, state and non-state, governmental and nongovernmental, reach a temporal convergence of expectations, making the political and analytical investments required to reimagine systems of global environmental governance in ways that reflect the transformative change that has occurred in humanity's relationship with earth systems? If anything, as the institutions of global environmental governance have matured, the number of actors relevant to realizing such change have multiplied, deepening temporal coordination dilemmas. The growing interrelatedness and complexity of cooperation problems, combined perhaps with the depth and breadth of the transformation required, means that the transaction costs associated with such institutional change are significant. The ongoing challenge of addressing the coronavirus pandemic has also made it increasingly difficult to generate the widespread focus needed for convergent expectations to emerge in global environmental governance. Coordination dilemmas are severe.

Yet, this book has shown that significant changes to international institutions can and do occur. My research calls attention to the importance of temporality and, in particular, Temporal Focal Points, in achieving significant change in international institutions. Cooperation takes place in real time, and it is by no means guaranteed that the "stars will align," enabling the far-reaching change required to tackle urgent world environmental problems. When expectations

[36] Biermann et al. 2012.

converge, however, principled actors must be ready to seize the opportunity. Indeed, the case studies examined in this book have shown the crucial importance of actors able to calibrate their behavior to capitalize on the structural forces at play within the wider international system, such as the availability of change incentives or coordination opportunities. We may be approaching a Temporal Focal Point now with the election of Joseph Biden in the United States, a sobering August 2021 Assessment Report from the IPCC, and widening recognition of significant, irreversible planetary impacts. Geopolitical competition may act as an accelerant of progress toward a no carbon future, as leading actors jockey for a dominant position in the future world energy order. We need to be ready to seize the moment.

Bibliography

Interviews

Beckerman, Professor Wilfred. June 11, 2010, Oxford, UK.

Björkbom, Lars. July 24, 2012, Stockholm, Sweden.

Brenton, Sir Anthony. July 19, 2012, Cambridge, UK.

Brown, Gordon S. April 3, 2013, via Telephone from Washington, DC, USA.

Bruce, Geoffrey. November 15, 2012, Ottawa, Canada.

Brundtland, Gro Harlem. August 17, 2021, via email from Oslo, Norway.

Buckley, James L. March 26, 2013, via Telephone from Sharon, USA.

Caballero, Paula. July 30, 2021, via email from Washington, DC, USA.

Dowdeswell, Elizabeth. September 7, 2021, via Zoom from Toronto, Canada.

Engfeldt, Lars-Göran. July 23, 2012, Visby, Sweden.

Gardner, Richard N. September 22, 2012, via Telephone from New York, USA.

Holdgate, Sir Martin. August 6, 2012, Kirkby Stephen, UK.

Koh, Tommy. August 21, 2017, via email from Singapore.

Long, Bill L. July 29, 2013, via Telephone from Washington, USA.

MacNeill, Dr. Jim. July 2, 2013, Ottawa, Canada.

MacNeill, Dr. Jim. August 18, 2010, Ottawa, Canada.

Munro, Bob. July 3, 2013, via Telephone from Nairobi, Kenya.

Runnalls, David. August 13, 2010, Ottawa, Canada.

Seaborn, J. Blair. April 18, 2013, Ottawa, Canada.

Strong, Maurice. August 20, 2010, Toronto, Canada.

Tickell, Sir Crispin. August 10, 2012, Oxford, UK.

Archival Materials

Harvard University Environmental Science and Public Policy Library, Cambridge, MA, USA.

Maurice F. Strong Papers

Peter S. Thacher Papers

Itamaraty, Government of Brazil, Brasília, Brazil.

Library and Archives Canada, Ottawa, Canada.
National Archives, London, UK.
Riksarkivet, Stockholm, Sweden.
Rockefeller Archive Center, Sleepy Hollow, NY, USA.
United Nations Archives and Record Management Section, New York, USA.

Public Documents

Brundtland, Gro Harlem. "Address to UNCTAD VII," Government of Norway (July 10, 1987a). www.regjeringen.no/upload/SMK/Vedlegg/Taler%20og%20artikler%20av%20tidligere%20statsministre/Gro%20Harlem%20Brundtland/1987/Address_to_UNCTAD_VII.pdf (accessed August 12, 2012).

"Statement at the 42nd Session of the General Assembly of the United Nations" (October 19, 1987b), Government of Norway. www.regjeringen.no/upload/SMK/Vedlegg/Taler%20og%20artikler%20av%20tidligere%20statsministre/Gro%20Harlem%20Brundtland/1987/Presentation_of_WCED_Report_to_UN_General_Assembly.pdf (accessed August 5, 2012).

Climate Action Tracker. "Paris Agreement Turning Point: Wave of Net Zero Targets Reduces Warming Estimate to 2.1°C in 2100, All Eyes on 2030 Targets" (December 2020). https://newclimate.org/wp-content/uploads/2020/12/CAT_2020-12-01_Briefing_GlobalUpdate_Paris5Years_Dec2020.pdf (accessed December 29, 2020).

"The Cocoyoc Declaration: Adopted by the Participants in the UNEP/UNCTAD Symposium on 'Patterns of Resource Use, Environment and Development Strategies.'" Cocoyoc, Mexico (October 8–12, 1974); International Union for Conservation of Nature and Natural Resources, *World Conservation Strategy: Living Resource Conservation for Sustainable Development* (IUCN, 1980); and The Brandt Commission, *North-South – A Programme for Survival,* (London: Pan Books, 1980).

Commonwealth Secretariat. *Sustainable Development: An Imperative for Environmental Protection* (London: Commonwealth Secretariat, 1991).

Department of External Affairs and International Trade. *From Stockholm to Rio 1972–1992 Report of the Seminar Convened by the Canadian Department of External Affairs and International Trade and the Department of Environment*" (December 8–9, 1991).

"Development and Environment: Report of the Panel of Experts Convened by the Secretary-General of the United Nations Conference on the Human Environment (Founex, Switzerland, 4–12 June 1971)" (1971)

United Nations Conference on the Human Environment: Development and Environment (Subject Area V), A/CONF.48.10, Annex I.

ECOSOC. "Statement by U.S. Ambassador Adlai Stevenson to the Thirty-Ninth Session of the Economic and Social Council" (June 9, 1965), ECOSOC Official Records, Doc. ESR.1375.

Intergovernmental Inter-sessional Preparatory Committee. "Institutional Arrangements for International Environmental Co-operation" (November 13, 1986), Doc. 1899e.

Intergovernmental Inter-sessional Preparatory Committee. "Environmental Perspective to the Year 2000 and Beyond," United Nations General Assembly, 42/186 (December 11, 1987).

Intergovernmental Panel on Climate Change. Climate Change 2021: The Physical Science Basis: Working Group I Contribution to the Sixth Assessment Report of the Intergovernmental Panel on Climate Change (August 7, 2021).

Intergovernmental Science-Policy Platform on Biodiversity and Ecosystem Services. IPBES Workshop Report on Biodiversity and Pandemics (2020). https://ipbes.net/pandemics (accessed August 14, 2021).

Johnson, Lyndon B. "Special Message to the Congress on Conservation and Restoration of Natural Beauty, February 8, 1965." Lyndon B. Johnson Presidential Library (1965). www.lbjlib.utexas.edu/johnson/archives.hom/speeches.hom/650208.asp (accessed June 26, 2013).

Kennedy, John F. "Address at 18th U.N. General Assembly, 20 September 1963," John F. Kennedy Presidential Library (1963). www.jfklibrary.org/Asset-Viewer/Archives/JFKPOF-046-041.aspx (accessed June 26, 2013).

"Letter Dated 20 March from the Permanent Representative of Mexico to the United Nations Office at Geneva Addressed to the Secretary-General of the UNCED (Tlatelolco Platform)," A/CONF.151/PC/L.30 (March 22, 1991). [Reproduced in: Netherlands Institute for the Law of the Sea, *International Organizations and the Law of the Sea: Documentary Yearbook* (Dordrecht: Martinus Nijhoff Publishers, 1993.]

Our World in Data. "SDG Tracker" (n.d.). https://sdg-tracker.org (accessed December 29, 2020).

Preparatory Committee for the United Nations Conference on Environment and Development, "Preparations for the United Nations Conference on Environment and Development on the Basis of General Assembly Resolution 44/228 and Taking into Account Any Other Relevant General Assembly Resolutions: Institutional Proposals – Report by the Secretary General of the Conference" (December 13, 1991), A/CONF.151/PC/102.

Preparatory Committee for the United Nations Conference on the Human Environment, Second Session (February 8–19, 1971), A/CONF.48/PC.8.

UN DESA. High Level Dialogue on Strengthening the Institutional Framework for Sustainable Development, Solo, Indonesia (July 19–21, 2011), Discussion paper.

UNCTAD. "Informal Encounter on International Trade and Environment" (February 28–March 1, 1991), UNCTAD/VIII/2, GE-91-50739.

UNCTAD. Environment and International Trade: Report of the Secretary-General of UNCTAD Submitted to the Secretary-General of the Conference (UNCED) Pursuant to General Assembly Resolution 45/210 (1992).

United Nations Environment Programme. "Declaration of the United Nations Conference on the Human Environment," United Nations Environment Programme Website (June 16, 1972). www.unep.org/Documents.Multilingual/Default.asp?documentid=97&articleid=1503 (accessed April 22, 2012).

United Nations Environment Programme. "Review of the Institutional Arrangements for International Environmental Co-operation" (February 16, 1976a), UNEP/GC/75.

United Nations Environment Programme. "Introductory Statement by the Executive Director: Four Years after Stockholm" (March 24, 1976b), UNEP/GC/L.36.

United Nations Environment Programme. "Report of the Governing Council on the Work of Its Ninth Session" (May 13–26, 1981), General Assembly, 36th Session, Suppl. No. 25, A/36/25.

United Nations Environment Programme. "Report of the Governing Council (Session of a Special Character and Tenth Session)" (1982), General Assembly, 37th Session, Suppl. No. 25, A/37/25.

United Nations Environment Programme. "'UNEP in the Next Decade' Executive Summary of the Seminar Deliberations – Note by the Executive Director" (March 25, 1986), Doc. 0174W.

United Nations Environment Programme. "Suggested Text of a Draft Resolution on Sustainable Development and the Environmental Perspective to the Year 2000 and Beyond for Submission by the Governing Council to the General Assembly" (1987) UNEP/GC.14/4/Add.7, Annex I.

United Nations Environment Programme. "Proceedings of the Governing Council at Its Sixteenth Session" (May 20–31, 1991), UNEP/GC.16/27.

United Nations Environment Programme. Emissions Gap Report (2020). www.unenvironment.org/emissions-gap-report-2020. (accessed January 5, 2021).

United Nations General Assembly. “Process of Preparation of the Environmental Perspective to the Year 2000 and Beyond,” Resolution 38/161 (December 19, 1983).

United Nations General Assembly. “Report of the World Commission on Environment and Development” (December 11, 1987), A/RES/42/187.

United Nations General Assembly Resolution 2398 (XXIII). “Problems of the Human Environment” (December 3, 1968).

United Nations General Assembly Resolution 2657 (XXV). “United Nations Conference on the Human Environment” (December 7, 1970).

United Nations General Assembly Resolution 2997 (XXVII). “International and Financial Arrangements for International Environmental Cooperation” (December 15, 1972).

United Nations General Assembly Resolution 35/74. “International Co-operation on the Field of the Environment” (December 5, 1980).

United Nations General Assembly Resolution 36/189. “Session of a Special Character of the Governing Council of the United Nations Environment Programme” (December 17, 1981).

United Nations General Assembly Resolution 44/228. “United Nations Conference on Environment and Development” (December 22, 1989), A/RES/44/228.

United Nations General Assembly Resolution 45/221. “United Nations Conference on Environment and Development” (December 21, 1990), A/RES/45/221.

United Nations General Assembly Resolution 53/242. “Report of the Secretary-General on Environment and Human Settlements” (August 10, 1999), A/RES/53/242.

United Nations General Assembly Resolution 55/199. “Ten-Year Review of Progress Achieved in the Implementation of the Outcome of the United Nations Conference on Environment and Development” (February 5, 2001), A/RES/55/199.

United Nations General Assembly Resolution 66/288. “The Future We Want” (September 11, 2012), A/RES/66/288.

United Nations General Assembly Resolution 67/213. “Report of the Governing Council of the United Nations Environment Programme on Its Twelfth Special Session and the Implementation of Section IV.C, Entitled ‘Environmental Pillar in the Context of Sustainable Development’, of the Outcome Document of the United Nations Conference on Sustainable Development” (December 21, 2012), A/RES/67/213.

United Nations General Assembly Resolution 67/290. “Format and Organizational Aspects of the High-Level Forum on Sustainable Development” (August 23, 2013), A/RES/67/290.

United Nations General Assembly Resolution 70/1. "Transforming Our World: The 2030 Agenda for Sustainable Development" (October 21, 2015), A/RES/70/1.

United Nations General Assembly Resolution 71/313. "Work of the Statistical Commission Pertaining to the 2030 Agenda for Sustainable Development" (July 10, 2017), A/RES/71/313.

United Nations General Assembly Resolution 72/277. "Towards a Global Pact for the Environment" (May 14, 2018), A/RES/72/277.

United Nations Information Service. Economic Commission for Africa – Addis Ababa Press Release (September 2, 1971), ECA/545,.

United Nations Press Release. "General Assembly Hears World Leaders Support Recommendations of World Commission on Environment and Development, Statements to Continue at Meeting This Afternoon" (October 19, 1987), GA/7529.

United States Congress – House Committee on Foreign Affairs, Subcommittee on Human Rights and International Organizations. Review of the Global Environment 10 Years after Stockholm: Hearings before the Subcommittee on Human Rights and International Organizations of the Committee on Foreign Affairs, House of Representatives, Ninety-Seventh Congress, Second Session, March 30, April 1 and 20, 1982 (Washington, DC: U.S. Government Printing Office, 1982).

United States Council on Environmental Quality and Department of State. Global Future: Time to Act – Report to the President on Global Resources, Environment and Population (Washington, DC: U.S. Government Printing Office, 1981).

United States Department of State. Foreign Relations of the United States, vol. E–1 (2005), Document on Global Issues, 1969–1972. http://history.state.gov/historicaldocuments/frus1969-76ve01 (accessed September 30, 2013).

United States Senate. Beyond the Earth Summit: Developing an Energy-Efficient World: Hearing before the Committee on Governmental Affairs, United States Senate, One Hundred Second Congress, Second Session, May 21, 1992 (Washington, DC: U.S. Government Printing Office, 1993).

Books and Secondary Sources

Abbott, Kenneth W. and Duncan Snidal. "Why States Act through Formal International Organizations," *Journal of Conflict Resolution*, vol. 42, no. 1 (February 1998), pp. 3–32.

Abbott, Kenneth W. and Steven Bernstein. "The High-Level Political Forum on Sustainable Development: Orchestration by Default and Design," *Global Policy*, vol. 6, no. 3 (September 2015), pp. 222–233.

Adams, William M. *Green Development: Environment and Sustainability in the Third World*, 2nd ed. (London: Routledge, 2001).

Agarwal, Anil. "The North-South Perspective: Alienation or Interdependence?," *Ambio*, vol. 19, no. 2 (April 1990), pp. 94–96.

Ågesta Group AB. Twenty Years after Stockholm, 1972–1992: A Report on the Implementation of the Stockholm Action Plan and on Priorities and Institutional Arrangements for the 1980s. A Parliamentary View of the State of the World Environment: The Results of a Survey Carried out by the Secretariat of the International Parliamentary Conference on the Environment (Berlin: Erich Schmidt Verlag GmbH, 1982).

American Geophysics Union. "New Studies Highlight Challenge of Meeting Paris Agreement Climate Goals," *Science Daily* (April 23, 2019). www.sciencedaily.com/releases/2019/04/190423114029.htm (accessed December 29, 2020).

Andersen, Inger. Conversation with Maria Ivanova, "Towards UNEP at 50: A Conversation with Inger Andersen" (May 7, 2021). www.youtube.com/watch?v=mp5ynMHouYo (accessed July 17, 2021).

Andresen, Steinar. "The Effectiveness of UN Environmental Institutions," *International Environmental Agreements*, vol. 7, no. 4 (2007), pp. 317–336.

de Araujo Castro, Joao Augusto. "Environment and Development: The Case of the Developing Countries," *International Institutions and the Environmental Crisis, International Organization* vol. 26, no. 2 (Spring 1972), pp. 401–416.

D'Arge, Ralph C. and Allen V. Kneese, "Environmental Quality and International Trade," *International Organization*, vol. 26, no. 2 (Spring 1972), pp. 419–465.

Arthur, W. Brian. *Increasing Returns and Path Dependence in the Economy* (Ann Arbor: University of Michigan Press, 1994).

Avant, Deborah and Oliver Westerwinter, "Introduction: Networks and Transnational Security Governance," in Deborah Avant and Oliver Westerwinter (eds.), *The New Power Politics: Networks and Transnational Security* (Oxford: Oxford University Press, 2016), pp. 1–18.

Axelrod, Robert. *The Evolution of Cooperation* (New York: Basic Books, 1981).

Ban, Ki-Moon. *Resolved: Uniting Nations in a Divided World* (New York: Columbia University Press, 2021).

Barabási, Albert-László. *Bursts: The Hidden Patterns behind Everything We Do, from Your E-Mail to Bloody Crusades* (New York: Penguin, 2011).

Barnett, Michael N. and Martha Finnemore. "The Politics, Power, and Pathologies of International Organizations," *International Organization*, vol. 53, no. 4 (Autumn 1999), pp. 699–732.

Barros, Antonio Teixeira de. "Brazil's Discourse on the Environment in the International Arena," *Contexto International*, vol. 39, no. 2 (2017), pp. 421–442.

Bartson, R. P. *Modern Diplomacy*, 5th ed. (London: Routledge, 2019).

Baumgartner, Frank R., Christian Breunig, Christoffer Green-Pedersen, et al. "Punctuated Equilibrium in Comparative Perspective," *American Journal of Political Science*, vol. 53, no. 3 (2009), pp. 603–620.

Benitez, Helen. "Success for a Sub-regional 'ASEAN' Plan," *Uniterra*, vol. 2 (1982), p. 25.

Bennett, Andrew and Alexander L. George. "Process Tracing in Case Study Research," Paper presented at the MacArthur Foundation Workshop on Case Study Methods, Belfer Center for Science and International Affairs (BCSIA), October 1997, Harvard University, Cambridge, MA.

Berenstain, Stan and Jan Berenstain. *Don't Pollute (Anymore)* (New York: Random House, 1991).

Bergesen, Helge Ole and Anne Kristin Sydnes. "Protection of the Global Climate: Ecological Utopia or Just a Long Way to Go?," in H. O. Bergesen, M. Norderhaug, and G. Parmann (eds.), *Green Globe Yearbook* (Oxford: Oxford University Press, 1992), pp. 35–48.

Bergesen, Helge Ole, Magnar Norderhaug, and Georg Parmann (eds.). *Green Globe Yearbook 1992* (Oxford: Oxford University Press, 1992).

Bernstein, Steven. *The Compromise of Liberal Environmentalism* (New York: Columbia University Press, 2001).

"Rio + 20: Sustainable Development in a Time of Multilateral Decline," *Global Environmental Politics*, vol. 13, no. 4 (November 2013), pp. 12–21.

Berridge, G. R. *Diplomacy: Theory and Practice* (Houndmills: Palgrave, 2002).

Berruga, Enrique and Peter Maurer. "Informal Consultations on Environmental Activities Co-chair Summary," in Lydia Swart and Estelle Perry (eds.), *Global Environmental Governance: Perspectives on the Current Debate* (New York: Center for UN Reform Education, 2007), pp. 16–25.

Biden, Joseph, Jr. "The Power of America's Example: The Biden Plan for Leading the Democratic World to Meet the Challenges of the 21st Century" (2021). https://joebiden.com/americanleadership/ (accessed January 5, 2021).

Biermann, Frank. "The Case for a World Environment Organization," *Science*, vol. 42, no. 9 (2000), pp. 22–31.

Earth System Governance: World Politics in the Anthropocene (Cambridge, MA: MIT Press, 2014).

Biermann, Frank, Kenneth Abbott, Steinar Andresen, et al. "Navigating the Anthropocene: Improving Earth System Governance," *Science*, vol. 335, no. 6074 (March 16, 2012), pp. 1306–1307.

Biermann, Frank, Philipp Pattberg, Harro van Asselt, and Fariborz Zelli. "The Fragmentation of Global Governance Architectures: A Framework for Analysis," *Global Environmental Politics*, vol. 9, no. 4 (November 2009), pp. 14–40.

Biermann, Frank and Rakhyun Kim. *Architectures of Earth System Governance: Institutional Complexity and Structural Transformation* (Cambridge: Cambridge University Press, 2020).

Binswanger, Hans. "Fiscal and Legal Incentives with Environmental Effects on the Brazilian Amazon," World Bank Working Paper (Washington: World Bank, 1987).

Black, Richard. "Key Powers Reach Compromise at Climate Summit," *BBC*, December 19, 2009. http://news.bbc.co.uk/2/hi/europe/8421935.stm (accessed August 10, 2021).

Blyth, Mark. "Powering, Puzzling, or Persuading? The Mechanisms of Building Institutional Orders," *International Studies Quarterly*, vol. 51, no. 4 (December 2007), pp. 761–777.

Borowy, Iris. *Defining Sustainable Development for Our Common Future: A History of the World Commission on Environment and Development (Brundtland Commission)* (London: Routledge, 2014).

Boughton, James M. "It Is Finally Time for a New Bretton Woods," Centre for International Governance Innovation (June 22, 2020). www.cigionline.org/articles/it-finally-time-new-bretton-woods (accessed July 26, 2020).

Brenton, Tony. *The Greening of Machiavelli: The Evolution of International Environmental Politics* (London: Earthscan, 1994).

Bruce, Geoffrey. "Canada's Road from Stockholm to Rio," in: Environment Canada (ed.), *From Stockholm to Rio: Summary Report of the Seminar Convened by the Canadian Department of External Affairs and International Tarade and the Department of Environment*, Willson House, Meech Lake, Quebec, December 8–9, 1991.

Brundtland, Gro Harlem. *Madam Prime Minister: A Life in Power and Politics* (New York: Farrar, Straus and Giroux, 2005).

Buckley, James L. "International Environmental Issues," *Department of State Bulletin*, vol. 57 (June 1982).

Bulmer-Thomas, Victor. "Life after Debt: The New Economic Trajectory in Latin America," Institute of Latin American Studies, University of Environmental Politics (London: Earthscan, 1992).

Büthe, Tim. "Historical Institutionalism and Institutional Development in the EU: The Development of Supranational Authority over Government Subsidies (State Aid)," in Thomas Rixen and Lora Anne Viola (eds.), *Historical Institutionalism and International Relations* (Oxford: Oxford University Press, 2016), pp. 37–67.

Caballero, Paula. "A Short History of the SDGs," Impakter (September 20, 2016). https://impakter.com/short-history-sdgs/ (accessed December 27, 2020).

Caldwell, Lynton Keith. *International Environmental Policy: Emergence and Dimensions*, 2nd ed. (Durham, NC: Duke University Press, 1990).

Canadian Environmental Advisory Council. *Canada and Sustainable Development* (Ottawa: Ministry of Supply and Services Canada, 1987).

Calvert, Randall L. "Rational Actors, Equilibrium, and Social Institutions," in Jack Knight and Itai Sened (eds.), *Explaining Social Institutions* (Ann Arbor: University of Michigan Press, 1998), pp. 57–94.

Capoccia, Giovanni. "Critical Junctures," in: Orfeo Fioretos, Tulia G. Falleti, and Adam Sheingate (eds.), *The Oxford Handbook of Historical Institutionalism* (Oxford: Oxford University Press, 2016), pp. 89–106.

Capoccia, Giovanni and Daniel Kelemen, "The Study of Critical Junctures: Theory, Narratives, and Counterfactuals in Historical Institutionalism," *World Politics*, vol. 59, no. 3 (2007), pp. 341–369.

Carlsson, Hans and Eric van Damme, "Global Games and Equilibrium Selection," *Econometrica*, vol. 61, no. 5 (September 1993), pp. 989–1018.

Carney, Mark, "Completing the G20 Reform Agenda for Strengthening Over-the-Counter Derivatives Markets," *Financial Stability Review* (April 2013), www.banque-france.fr/sites/default/files/medias/documents/financial-stability-review-17_2013-04.pdf (accessed January 7, 2022).

Carpenter, R. Charli, "Vetting the Advocacy Agenda: Network Centrality and the Paradox of Weapons Norms," *International Organization*, vol. 65, no. 1 (Winter 2011), pp. 69–102.

Carson, Rachel. *Silent Spring* (Boston: Houghton Mifflin, 1962).

Castells, Manuel. *Networks of Outrage and Hope: Social Movements in the Internet Age* (New York: Polity, 2012).

Chan, Nicholas. "Climate Contributions and the Paris Agreement: Fairness and Equity in a Bottom-Up Architecture," *Ethics & International Affairs*, vol. 30, no. 3 (2016), pp. 291–301.

Chapnick, Adam. *The Middle Power Project: Canada and the Founding of the United Nations* (Vancouver: University of British Columbia Press, 2005).

Chasek, Pamela S. and David L. Downie, *Global Environmental Politics*, 8th ed. (New York: Routledge, 2021).

Chirac, Jacques. "Speech to the Plenary Session of the World Summit on Sustainable Development," Johannesburg (September 2, 2002) www.jacqueschirac-asso.fr/archives-elysee.fr/elysee/elysee.fr/nglaise/speeches_and_documents/2002-2001/fi005004.html (accessed July 9, 2021).

Choucri, Nazali. "Population, Resources, and Technology: Political Implications of the Environmental Crisis," *International Institutions and the Environmental Crisis, International Organization* vol. 26, no. 2 (Spring 1972), pp. 175–212.

Clarke, Robin and Lloyd Timberlake, *Stockholm Plus Ten: Promises, Promises? The Decade since the 1972 UN Environment Conference* (London: Earthscan, 1982).

Collier, Ruth Berins and David Collier, *Shaping the Political Arena: Critical Junctures, the Labor Movement, and Regime Dynamics in Latin America* (Princeton, NJ: Princeton University Press, 1991).

Commoner, Barry. *The Closing Circle: Nature, Man, and Technology* (New York: Knopf, 1971).

Cooney, John "EEC Looks for Ecology Aid," *The Irish Times* (April 14, 1982), p. 8.

Corea, Gamani. *My Memoirs* (Ratmalana: Gamani Corea Foundation, 2008).

Corrêa do Lago, André Aranha. *Stockholm, Rio, Johannesburg: Brazil and the Three United Nations Conferences on the Environment* (Brasília: Ministry of External Relations, 2009).

Crocitti, John J. and Monique Vallance. *Brazil Today* (Santa Barbara: ABC-CLIO, 2012).

Culpepper, Roy. "Managing the Global Commons: The Economic Setting and Financial Options," in Environment Canada (ed.), *From Stockholm to Rio: Summary Report of the Seminar Convened by the Canadian Department of External Affairs and International Tarade and the Department of Environment*, Willson House, Meech Lake, Quebec, December 8–9, 1991.

DiMaggio, Paul J. and Walter W. Powell, "The Iron Cage Revisited: Institutional Isomorphism and Collective Rationality in Organizational Fields," *American Sociological Review*, vol. 48, no. 2 (April 1983), pp. 147–160.

Dodds, Felix and Anita Nayar, "Rio+20: A New Beginning," *United Nations Environment Programme: Perspectives*, no. 8 (June 20, 2015), pp. 1–20.

Dodds, Felix and Michael Strauss, with Maurice Strong, *Only One Earth: The Long Road via Rio to Sustainable Development* (London: Routledge, 2012).

Dryzek, J. S. "Institutions for the Anthropocene: Governance in a Changing Earth System," *British Journal of Political Science*, vol. 46, no. 4 (2016), pp. 937–956.

Eckholm, Erik P. *Down to Earth: Environment and Human Needs* (New York: Norton, 1982).

Eckstein, Harry. "Case Studies in Political Science," in Fred Greenstein and Nelson Polsby (eds.), *Handbook of Political Science*, vol. 7 (Reading, MA: Addison-Wesley, 1975), pp. 79–138.

Ecologist, *A Blueprint for Survival* (Middlesex: Penguin, 1972).

ENDA-Tiers Monde, Poverty and Environment in Africa: Which Way Ahead? Paper 21, December 1991 (London: IIED, 1991).

Ehrlich, Paul R. *The Population Bomb* (New York: Ballantine Books, 1968).

Eldredge, Niles and Stephen Jay Gould, "Punctuated Equilibria: An Alternative to Phyletic Gradualism," in Thomas J. M. Schopf (ed.), *Models in Paleobiology* (San Francisco, CA: Freeman, Cooper & Company, 1972), pp. 82–115.

Engfeldt, Lars-Göran. *From Stockholm to Johannesburg and Beyond: The Evolution of the International Systems for Sustainable Development Governance and Its Implications* (Stockholm: The Government Offices of Sweden, 2009).

Evans, Alex and David Steven, *Sustainable Development Goals – A Useful Outcome from Rio+20?* (New York: NYU Center for International Cooperation, 2012).

Fearon, James D. "Signaling versus the Balance of Power and Interests: An Empirical Test of a Crisis Bargaining Model," *The Journal of Conflict Resolution*, vol. 38, no. 2 (June 1994), pp. 236–269.

"Bargaining, Enforcement and International Cooperation," *International Organization*, vol. 52, no. 2 (Spring 1998), pp. 269–305.

Figueres, Christiana and Tom Rivett-Carnac, *The Stubborn Optimist's Guide to the Climate Crisis* (New York: Vintage, 2020).

Finnemore, Martha and Kathryn Sikkink, "International Norm Dynamics and Political Change," *International Organization*, vol. 52, no. 4 (Autumn 1998), pp. 887–917.

Fioretos, Orfeo. "Historical Institutionalism and International Relations," *International Organization*, vol. 65, no. 2 (Spring 2011), pp. 367–399.

Fioretos, Orfeo. (ed.). *International Politics and Institutions in Time* (Oxford: Oxford University Press, 2017).

Fisher, Roger, William Ury, and Bruce Patton, *Getting to Yes: Negotiating Agreement Without Giving In*, 2nd ed. (London: Penguin Books, 1991).

Fitzgerald, Mary Anne. "U.N. Environment Conference Ends Feeling Pinch of Reduced U.S. Role," *The Washington Post* (May 19, 1982) p. A19.

Fomerand, Jacques. "UN Conferences: Media Events or Genuine Diplomacy," *Global Governance*, vol. 2, no. 3 (September–December 1996), pp. 361–375.

Food and Agriculture Organization of the United Nations, *Sustainable Development and the Environment: FAO Policies and Actions, Stockholm 1972–Rio 1992* (Rome: FAO, 1992).

Fortune. "The Environment," *Fortune*, special edition (February 1970).

Founex Report, Environment and Development, Development and Environment: Report and Working Papers of a Panel of Experts Convened by the Secretary-General of the United Nations Conference on the Human Environment (Founex, Switzerland, June 4–12, 1971) (Paris: Mouton, 1972).

Pope Francis, Laudato Si: On Care for Our Common Home (2015). www.vatican.va/content/francesco/en/encyclicals/documents/papa-francesco_20150524_enciclica-laudato-si.html (accessed July 22, 2021).

Friends of the Earth. *The Stockholm Conference Only One Earth: An Introduction to the Politics of Survival* (London: Earth Island Limited, 1972).

Environmentally Sensitive Areas: Assessment and Recommendations (London: Friends of the Earth, 1992).

Gardner, Richard N. "The Role of the UN in Environmental Problems," *International Institutions and the Environmental Crisis, International Organization*, vol. 26, no. 2 (Spring 1972), pp. 237–354.

Garrett, Geoffrey and Barry R. Weingast, "Ideas, Interests, and Institutions: Constructing the European Community's Internal Market," in Judith Goldstein and Robert O. Keohane (eds.), *Ideas & Foreign Policy: Beliefs, Institutions, and Political Change* (Ithaca, NY: Cornell University Press, 1993), pp. 173–206.

Gellman, Barton. "Slaughter in Racak Changed Kosovo Policy, *The Washington Post* (April 18, 1999), p. A1.

General Agreement on Tariffs and Trade. "Industrial Pollution Control and International Trade," in Development and Environment: Report and Working Papers of a Panel of Experts Convened by the Secretary-General of the United Nations Conference on the Human Environment (Founex, Switzerland, June 4–12, 1971) (Paris: Mouton, 1972).

George, Alexander L. and Andrew Bennett. *Case Studies and Theory Development in the Social Sciences* (Cambridge, MA: MIT Press, 2005).

Germain, G., A. Simon, J. Arsenault, et al. Quebec's Multi-Party Observatory on Zoonoses and Adaptation to Climate Change, "Quebec's Multi-Party Observatory on Zoonoses and Adaptation to

Climate Change," *The Canadian Communicable Disease Report*, vol. 45, no. 5 (May 2019), pp. 143–148.

Gerschewski, Johannes, "Explanations of Institutional Change: Reflecting on a "Missing Diagonal," *American Political Science Review*, vol. 115, no. 1 (2021), pp. 218–233.

Gersick, Connie J. G. "Time and Transitions in Work Teams: Toward a New Model of Group Development," *The Academy of Management Journal*, vol. 31, no. 1 (March 1988), pp. 9–41.

"Making Time Predictable: Transitions in Task Groups," *The Academy of Management Journal*, vol. 32, no. 2 (June 1989), pp. 274–309.

Gilpin, Robert. *War and Change in World Politics* (Cambridge: Cambridge University Press, 1981).

Goldsmith, Edward. "The Retreat from Stockholm," *The Ecologist*, vol. 12, no. 3 (May–June 1982), pp. 98–100.

Gorsuch, Anne. "Eight Principles for Sound Environmental Development," *Uniterra* vol. 2 (1982).

Gotlieb, Allan. *The Washington Diaries, 1981–1989* (Toronto: McClelland & Stewart, 2006).

Grief, Avner and David D. Laitin, "A Theory of Endogenous Institutional Change," *American Political Science Review*, vol. 98, no. 4 (November 2004), pp. 633–652.

Gruber, Lloyd, *Ruling the World: Power Politics and the Rise of Supranational Institutions* (Princeton, NJ: Princeton University Press, 2000).

Grunberg, Carole A. "Report on the United Nations Environment Programme's Session of a Special Character, May 10–18, 1982, Nairobi, Kenya," United States Congress, Review of the Global Environment 10 Years after Stockholm.

Guterres, António. "Secretary-General's Address at Columbia University: 'The State of the Planet'." (December 2, 2020). www.un.org/sg/en/content/sg/speeches/2020-12-02/address-columbia-university-the-state-of-the-planet (accessed July 17, 2021).

Haas, Peter M. "Banning Chlorofluorocarbons: Epistemic Community Efforts to Protect Stratospheric Ozone, *International Organization*, vol. 46, no. 1 (Winter 1992a), pp. 187–224.

"Introduction: Epistemic Communities and International Policy Coordination," *International Organization*, vol. 46, no. 1 (Winter 1992b), 1–35.

"UN Conferences and Constructivist Governance of the Environment," *Global Governance*, vol. 8, no. 1 (January–March 2002), pp. 73–91.

Haigh, Nigel. "The European Community and International Environmental Policy, in Andrew Hurrell and Benedict Kingsbury (eds.), *The*

International Politics of the Environment: Actors, Interests, and Institutions (Oxford: Oxford University Press, 1992), pp. 228–249.

Hall, Nina. *Displacement, Development, and Climate Change: International Organizations Moving beyond Their Mandates* (Abington: Routledge 2016).

Halle, Mark. "Life after Rio: A Commentary by Mark Halle, IISD," International Institute for Sustainable Development (June 2012). www.iisd.org/system/files/publications/com_life_after_rio.pdf (accessed August 1, 2021).

Hampson, Fen Osler, with Michael Hart. *Multilateral Negotiations: Lessons from Arms Control, Trade, and the Environment* (Baltimore: Johns Hopkins University Press, 1999).

Hanrieder, Tine and Michael Zürn. "Reactive Sequences in Global Health Governance," in Orfeo Fioretos (ed.), *International Politics and Institutions in Time* (Oxford: Oxford University Press, 2017).

ul Haq, Mahbub. *The Poverty Curtain: Choices for the Third World* (New York: Columbia University Press, 1976).

Harsanyi, John C. "Games with Randomly Distributed Payoffs: A New Rationale for Mixed-Strategy Equilibrium Points," *International Journal od Game Theory*, vol. 2 (1973) pp. 1–23.

Harvey, Fiona. "Trillions of Dollars Spent on Covid Recovery in Ways that Harm Environment," *The Guardian* (July 15, 2021). www.theguardian.com/business/2021/jul/15/trillions-of-dollars-spent-on-covid-recovery-in-ways-that-harm-environment (accessed August 16, 2021).

Hassner, Ron E. "Sacred Time and Conflict Initiation," *Security Studies*, vol. 20, no. 4 (2011), pp. 491–520.

Hay, Colin and Daniel Wincott, "Structure, Agency and Historical Institutionalism," *Political Studies*, vol. XLVI (1998), pp. 951–957.

Heath, Ryan, Karl Mathiesen, Zack Colman, and Kalina Oroschakoff, "World Pushes Ahead with Climate Action as U.S. Remains in Limbo," *Politio* (November 17, 2020). www.politico.com/news/2020/11/17/global-summit-climate-action-biden-436833 (accessed January 6, 2021).

Helleiner, Eric. "Incremental Origins of Bretton Woods," in: Orfeo Fioretos (ed.), *International Politics and Institutions in Time* (Oxford: Oxford University Press, 2017).

Herter Jr., Christian A. and Jill E. Binder. *The Role of the Secretariat in Multilateral Negotiation: The Case of Maurice Strong and the 1972 UN Conference on the Human Environment* (Washington, DC: Johns Hopkins Foreign Policy Institute, 1993).

Hickel, Jason. "The World's Sustainable Development Goals Aren't Sustainable," *Foreign Policy* (September 30, 2020). https://

foreignpolicy.com/2020/09/30/the-worlds-sustainable-development-goals-arent-sustainable/ (accessed December 29, 2020).

Holdgate, Martin W. *A Perspective of Environment Pollution* (Cambridge: Cambridge University Press, 1979).

Penguins and Mandarins: The Memoirs of Martin Holdgate (Spennymoor: The Memoir Club, 2003).

Holdgate, Martin W., Mohammed Kassas, and Gilbert F. White. *The World Environment 1972–1982: A Report by the United Nations Environment Programme* (Dublin: Tycooly International Publishing, 1982).

Hom, Andrew R. *International Relations and the Problem of Time* (Oxford: Oxford University Press, 2020).

Hopgood, Stephen. *American Foreign Environment al Policy and the Power of the State* (Oxford: Oxford University Press, 1998).

Howard, Brian Clark. "Rio+20 Brings Hope and Solutions Despite Weak Talk," *National Geographic* (June 21, 2012), www.nationalgeographic.com/science/article/120621-rio-20-hope-solutions-official-talks#close (accessed January 7, 2022).

Hurrell, Andrew. "The Politics of Amazonian Deforestation," *Journal of Latin American Studies*, vol. 23, no. 1 (1991), pp. 197–215.

"Brazil and Amazonian Deforestation," in Andrew Hurrell and Benedict Kingsbury (ed.), *The International Politics of the Environment: Actors, Interests, and Institutions* (Oxford: Oxford University Press, 1992), pp. 398–429.

Hurrell, Andrew and Benedict Kingsbury (eds.), *The International Politics of the Environment: Actors, Interests, and Institutions* (Oxford: Oxford University Press, 1992).

Ikenberry, G. John. *After Victory: Institutions, Strategic Restraint, and the Rebuilding of Order after Major Wars* (Princeton, NJ: Princeton University Press, 2000).

Independent Commission on Disarmament and Security Issues. *Common Security: A Blueprint for Survival* (New York: Simon & Schuster, 1982).

Independent Commission on International Development Issues. *North-South: A Programme for Survival* (New York: Pan Books, 1980).

Inomata, Tadanori. "Management Review of Environmental Governance within the United Nations System," JIU/REP/2008/3(Geneva: Joint Inspection Unit, 2008).

Institute of Latin American Studies. *The Debt Crisis in Latin America* (Stockholm: Latinamerika-institute, 1986).

International Council of Scientific Unions' Scientific Committee on Problems of the Environment and International Union for Pure and Applied

Biophysics. *Effects of Increased Ultraviolet Radiation on Biological Systems: Proceedings of a Workshop* (Paris: SCOPE, 1992).

International Institute for Sustainable Development (IISD). "Monday Evening, 23 March 1992," *Earth Negotiations Bulletin*, vol. 1, no. 18 (March 25, 1992a). www.iisd.ca/download/asc/enb0118.txt (accessed September 1, 2013).

"Monday Evening, 30 March 1992," *Earth Negotiations Bulletin*, vol. 1, no. 23 (March 30, 1992b). www.iisd.ca/download/asc/enb0123.txt (accessed September 4, 2013).

"Saturday Evening, 28 March 1992," *Earth Negotiations Bulletin*, vol. 1, no. 22 (March 31, 1992c). www.iisd.ca/download/asc/enb0122.txt (accessed September 4, 2013).

"70th Meeting of the Plenary," *Earth Negotiations Bulletin*, vol. 1, no. 26 (April 3, 1992d). www.iisd.ca/download/asc/enb0126.txt (accessed September 4, 2013).

"UNCED Highlights, 5–6 June 1992: The Contact Groups," *Earth Negotiations Bulletin*, vol. 2, no. 6 (June 7, 1992e). www.iisd.ca/vol02/0206002e.html (accessed September 5, 2013).

"UNCED Highlights, 8 June 1992: Main Committee Contact Groups," *Earth Negotiations Bulletin*, vol. 2, no. 7 (June 9, 1992f). www.iisd.ca/vol02/0207001e.html (accessed September 5, 2013).

"Summary Report, 23 April–4 May 2012: 2nd Round of UNCSD Informal Informal Consultations," *Earth Negotiations Bulletin*, vol. 27, no. 35 (2012a), p. 7. https://enb.iisd.org/events/2nd-round-uncsd-informal-informal-consultations/summary-report-23-april-4-may-2012 (accessed July 9, 2021).

"Summary Report, 29 May–2 June 2012: 3rd Round of UNCSD Informal Informal Consultations," *Earth Negotiations Bulletin*, vol. 27, no. 40 (June 5, 2012b), p. 7. https://enb.iisd.org/events/3rd-round-uncsd-informal-informal-consultations/summary-report-29-may-2-june-2012 (accessed July 9, 2021).

"Governments Begin Negotiating 2022 Political Declaration on Environment" (August 25, 2020a). https://sdg.iisd.org/news/governments-begin-negotiating-2022-political-declaration-on-environment/ (accessed December 29, 2020).

"Summary of the Climate Ambition Summit 2020: 12 December 2020," *Earth Negotiations Bulletin* (December 15, 2020b). https://enb.iisd.org/vol12/enb12778e.html (accessed January 5, 2021).

"HLPF Reviews Progress towards Nine SDGs" (July 15, 2021). https://sdg.iisd.org/news/hlpf-reviews-progress-towards-nine-sdgs/ (accessed December 26, 2021).

International Institute for Sustainable Development Reporting Services. "Summary of the United Nations Conference on Sustainable

Development: 13–22 June 2012," *Earth Negotiations Bulletin*, vol. 27, no. 51, Monday (June 25, 2012).

International Soil Reference and Information Centre. *World Status of Human-Induced Soil Degradation* (Wageningen: ISRIC, 1990).

International Union for Conservation of Nature and Natural Resources. *World Conservation Strategy: Living Resource Conservation for Sustainable Development* (Gland: IUCN-UNEP-WWF, 1980).

"Nature-Based Solutions." (2021). www.iucn.org/theme/nature-based-solutions (accessed July 18, 2021).

Ivanova, Maria. "Designing the United Nations Environment Programme: A Story of Compromise and Confrontation," *International Environmental Agreements: Politics, Law and Economics* vol. 7, no. 3 (2007), pp. 337–361.

"UNEP in Global Environmental Governance: Design, Leadership, Location," *Global Environmental Politics*, vol. 10, no. 1 (February 2010), pp. 30–59.

"Institutional Design and UNEP Reform: Historical Insights on Form, Function and Financing," *International Affairs*, vol. 88, no. 3 (2012), pp. 565–584.

"The Contested Legacy of Rio+20," *Global Environmental Politics*, vol. 13, no. 4 (November 2013a), pp. 1–11.

"Reforming the Institutional Framework for Environment and Sustainable Development: Rio+20's Subtle but Significant Impact," *International Journal of Technology Management & Sustainable Development*, vol. 12, no. 3 (September 2013b), pp. 211–231.

The Untold Story of the World's Leading Environmental Institution (Cambridge: MIT Press, 2021).

Jebb, Gladwyn. *The Memoirs of Lord Gladwyn* (London: Weybright and Talley, 1972).

Johnson, Boris. "PM Climate Ambition Summit Opening Remarks: 12 December 2020," Prime Minister's Office (December 12, 2020). www.gov.uk/government/speeches/pm-climate-ambition-summit-opening-remarks-12-december-2020 (accessed January 5, 2021).

Johnson, Stanley. *UNEP: The First Forty Years: A Narrative* (Nairobi: United Nations Environment Programme, 2012).

Jones, Bryan D. "From There to Here: Punctuated Equilibrium to the General Punctuation Thesis to a Theory of Government Information Processing," *Policy Studies Journal*, vol. 40, no. 1 (2012), pp. 1–20.

Jones, Bryan D. and Frank R. Baumgartner. *The Politics of Attention: How Government Prioritizes Problems* (Chicago: University of Chicago Press, 2005).

Jupille, Joseph and James A. Caporaso. *Theories of Institutions* (Cambridge: Cambridge University Press, 2021).

Jupille, Joseph, Walter Mattli, and Duncan Snidal, *Institutional Choice and Global Commerce* (Cambridge: Cambridge University Press, 2013).

Kahler, Miles. "Networked Politics: Agency, Power, and Governance," in Miles Kahler (ed.), *Networked Politics: Agency, Power, and Governance* (Ithaca, NY: Cornell University Press, 2009), pp. 1–22.

Kamau, Macharia, Pamela Chasek, and David O'Connor, *Transforming Multilateral Diplomacy: The Inside Story of the Sustainable Development Goals* (London: Routledge, 2018).

Keck, Margaret E. and Kathryn Sikkink. *Activists beyond Borders* (Ithaca, NY: Cornell University Press, 1998).

Kennett, Wayland. "The Stockholm Conference on the Human Environment, *International Affairs*, vol. 48, no. 1 (1972), pp. 33–45.

Keohane, Robert O. "The Demand for International Regimes," *International Organization*, vol. 36, no. 2 (Spring 1982), pp. 325–355.

After Hegemony: Cooperation and Discord in the World Political Economy (Princeton, NJ: Princeton University Press, 1984).

"International Institutions: Two Approaches," *International Studies Quarterly*, vol. 32, no. 4 (December 1988), pp. 379–396.

Keohane, Robert O. and Lisa L. Martin, "The Promise of Institutionalist Theory," *International Security*, vol. 20, no. 1 (Summer 1995), pp. 39–51.

Keohane, Robert O. and Joseph S. Nye, *Power and Interdependence*, 2nd ed. (New York: Harper-Collins, 1989).

Keohane, Robert O. and David G. Victor, "The Regime Complex for Climate Change," *Perspectives on Politics*, vol. 9, no. 1 (March 2011), pp. 7–23.

King, Gary, Robert O. Keohane, and Sidney Verba. *Designing Social Inquiry* (Princeton, NJ: Princeton University Press, 1994).

King, Tom. "Population – An Immense Priority Problem," *Uniterra*, vol. 2 (1982).

Kingdon, John W. *Agendas, Alternatives, and Public Policies* (New York: Longman, 1984).

Kinsman, Jeremy. "History at a Juncture," *Policy* (June 17, 2020). www.policymagazine.ca/history-at-a-juncture/ (accessed December 26, 2021).

Kjellén, Bo. *A New Diplomacy of Sustainable Development: A Challenge of Global Change* (New York: Routledge, 2008).

Krasner, Stephen D. "State Power and the Structure of International Trade," *World Politics*, vol. 28, no. 3 (April 1976), pp. 317–347.

"Approaches to the State: Alternative Conceptions and Historical Dynamics," *Comparative Politics*, vol. 16, no. 2 (January 1984), pp. 223–246.

"Global Communications and National Power: Life on the Pareto Frontier," *World Politics*, vol. 43, no. 3 (April 1991), pp. 336–366.

Kruck, Andreas and Bernhard Zangl, "The Adjustment of International Institutions to Global Power Shifts: A Framework for Analysis," *Global Policy*, vol. 11, no. S3 (2020) pp. 5–16.

Lamb, Robert P. "Guest Editorial: An Interview with UNEP Chief," *Environmental Conservation*, vol. 9, no. 1 (Spring 1982), p. 1.

Larson, Eric, Chris Grieg, Jesse Jenkins et al. "Net Zero America: Potential Pathways, Infrastructure, and Impacts" (December 2020). https://environmenthalfcentury.princeton.edu/sites/g/files/toruqf331/files/2020-12/Princeton_NZA_Interim_Report_15_Dec_2020_FINAL.pdf (accessed December 26, 2021).

Lashof, D. and D. A. Tirpak. *Policy Options for Stabilising Global Climate*. 21P-2003. (Washington, DC: US Environmental Protection Agency, 1990).

Lax, David L. and James K. Sebenius. *The Manager as Negotiator: Bargaining for Cooperative and Competitive Gain* (New York: The Free Press, 1986).

Lazar, Nomi Claire. *Out of Joint: Power, Crisis, and the Rhetoric of Time* (New Haven, CT: Yale University Press, 2019).

Le Prestre, Philippe G. *The World Bank and the Environmental Challenge* (Selinsgrove: Susquehanna University Press, 1989).

Lean, Geoffrey and Marek Mayer, "Britain Moving towards Acid Rain Disaster," *The Observer* (June 13, 1982), p. 2.

Lee, James A. "Environmental Considerations in Development Finance," *International Institutions and the Environmental Crisis, International Organization*, vol. 26, no. 2 (Spring 1972), pp. 337–347.

Leonard, H. Jeffrey and David Morell. "Emergence of Environmental Concern in Developing Countries: A Political Perspective," *Stanford Journal of International Law*, vol. 17 (1981), pp. 281–314.

Levy, Jack S. "Prospect Theory, Rational Choice, and International Relations," *International Studies Quarterly*, vol. 41, no. 1 (1997), pp. 87–112.

Lewis, Paul. "Poverty Is Linked to Harm to Land," *New York Times*, (October 20, 1987), p. A5.

Libecap, G. D. *Contracting for Property Rights* (Cambridge: Cambridge University Press, 1989).

Litfin, Karen T. *Ozone Discourse: Science and Politics in Global Environmental Cooperation* (New York: Columbia University Press, 1995).

Lim, Stephen Ghee-Soon and J. Keith Murnighan, "Phases, Deadlines, and the Bargaining Process," *Organizational Behavior and Human Decision Processes*, vol. 58 (1994), pp. 153–171.

Lipscy, Phillip Y. *Renegotiating the World Order: Institutional Change in International Relations* (Cambridge: Cambridge University Press, 2017).

Long, Bill L. *International Environmental Issues and the OECD 1950–2000* (Paris: OECD, 2000).

Lundgren, Magnus, Theresa Squatrito, and Jonas Tallberg. "Stability and Change in International Policy-Making: A Punctuated Equilibrium Approach," *Review of International Organizations*, vol. 13, no. 4 (2018), pp. 547–572.

Lyons, Richard D. "Furor over Plan to Dump Nerve Gas," *New York Times* (August 9, 1970).

MacDonald, Gordon J. "International Institutions for Environmental Management," *International Institutions and the Environmental Crisis, International Organization*, vol. 26, no. 2 (Spring 1972), pp. 372–400.

MacMillan, Margaret. *Paris 1919: Six Months that Changed the World* (New York: Random House, 2003).

"The World after COVID-19: Margaret MacMillan on Covid-19 as a Turning Point in History," *The Economist* (May 9, 2020). www.economist.com/by-invitation/2020/05/09/margaret-macmillan-on-covid-19-as-a-turning-point-in-history (accessed July 25, 2020).

"Which Past Is Prologue? Heeding the Right Warnings from History," *Foreign Affairs*, vol. 99, no. 5 (September/October 2020).

MacNeill, Jim. "Environment and Development," *International Perspectives*, vol. 17, no. 3 (May/June 1988), pp. 3–8.

"From Controversy to Consensus: Building Global Agreement for Change," *Environmental Policy and Law*, vol. 37, no. 2–3 (2007), pp. 2–3.

Mahoney, James. "Path Dependence in Historical Sociology," *Theory and Society*, vol. 29, no. 4 (August 2000), pp. 507–548.

Mahoney, James and Kathleen Thelen, "A Theory of Gradual Institutional Change," in James Mahoney and Kathleen Thelen (eds.), *Explaining Institutional Change: Ambiguity, Agency, and Power* (New York: Cambridge University Press, 2010), pp. 1–37.

Manulak, Michael W. "Canada and the Kosovo Crisis: A Golden Moment in Canadian Foreign Policy?" *International Journal*, vol. 64, no. 2 (2009), pp. 565–581.

"Developing World Environmental Cooperation: The Founex Seminar and the Stockholm Conference," in Wolfram Kaiser and John-Henrik Meyer (eds.), *International Organizations and Environmental Protection in the Global Twentieth Century: Actors, Agenda-Setting and Institution-Building* (New York: Berghahn Books, 2016), pp. 103–127.

"Leading by Design: Informal Influence and International Secretariats," *Review of International Organizations*, vol. 12, no. 4 (2017), pp. 497–522.

"A Bird in the Hand: Temporal Focal Points and Change in International Institutions," *Review of International Organizations*, vol. 15, no. 1 (2020), pp. 1–27.

"The Networked Diplomacy of Informal International Institutions: The Case of the Proliferation Security Initiative," *Global Governance*, vol. 27, no. 3 (Fall 2021a), pp. 410–432.

"Justin Trudeau and Institutional Informality: Canada's Response to the COVID-19 Crisis," in David Carment and Richard Nimijean (eds.), *Canada Among Nations 2020: Divided in a Dangerous World* (London: Palgrave-Macmillan, 2021b), pp. 211–232.

Manulak, Michael W. and Duncan Snidal, "Hierarchy Plus Networks in Global Governance: The Supply of Informal International Governance," in Michael Barnett, Jon Pevehouse, and Kal Raustiala (eds.), *Global Governance in a World of Change* (Cambridge: Cambridge University Press, 2021), pp. 182–213.

March, James G. and Johan P. Olsen, *Rediscovering Institutions: The Organizational Basis of Politics* (New York: The Free Press, 1989).

Martin, Lisa L. "Interests, Power, and Multilateralism," *International Organization*, vol. 46, no. 4 (Autumn 1992), pp. 765–792.

Marx, Eric, "Rio+20: Earth Summit – Expecting the World? Disappointment Inevitable," *Reuters* (June 11, 2012). www.reutersevents.com/sustainability/environment/rio20-earth-summit-expecting-world-disappointment-inevitable (accessed July 29, 2021).

Mattli, Walter and Ngaire Woods. *The Politics of Global Regulation* (Princeton, NJ: Princeton University Press, 2009).

McChrystal, Stanley, with Tantum Collins, David Silverman, and Chris Fussell. *Team of Teams: New Rules of Engagement for a Complex World* (New York: Penguin, 2015).

McCormick, John. *Reclaiming Paradise: The Global Environmental Movement* (Bloomington: Indiana University Press, 1991).

McDonald, John W. "Global Environmental Negotiations: The 1972 Stockholm Conference and Lessons for the Future," Paper for the *American Academy of Diplomacy and the Paul H. Nitze School of Advanced International Studies*, Johns Hopkins University (December 13, 1989).

McLaughlin, Kathleen. Comments on World Resource Institute Panel, "ADVISORY: Press Call on 5-Year Anniversary of the Paris Agreement and Expectations for the Climate Ambition Summit" (December 3, 2020). www.wri.org/news/2020/12/advisory-press-call-5-year-anniversary-paris-agreement-and-expectations-climate (accessed January 6, 2021).

Meadows, Dennis L. and Jorgen Randers, "Adding the Time Dimension to Environmental Policy," *International Organization* vol. 26, no. 2 (Spring 1972), pp. 213–233.

Meadows, Donella H., Dennis L. Meadows, Jørgen Randers, et al. *The Limits to Growth: A Report for the Club of Rome's Project on the Predicament of Mankind* (New York: Universe Books, 1972).

Meyer, John M. and Brian Rowan. "Institutionalized Organizations: Formal Structure as Myth and Ceremony," *American Journal of Sociology*, vol. 83, no. 2 (September 1977).

Meyer, John W. and W. R. Scott. *Organizational Environments: Ritual and Reality* (Beverly Hills: Sage, 1983), pp. 340–363.

Moi, Daniel arap. "Let's Take Our Coats Off and Get Honourably Dirty," *Uniterra*, vol. 2 (1982).

Moravcsik, Andrew. "Taking Preferences Seriously: A Liberal Theory of International Politics," *International Organization*, vol. 51, no. 4 (Autumn 1997), pp. 513–553.

"A New Statecraft? Supranational Entrepreneurs and International Cooperation," *International Organization*, vol. 53, no. 2 (Spring 1999), pp. 267–306.

"The New Liberalism," in Christian Reus-Smit and Duncan Snidal (eds.), *The Oxford Handbook of International Relations* (Oxford: Oxford University Press, 2008), pp. 234–254.

Munro, Bob. "Choosing Our Common Future," in Environment Canada (ed.), *From Stockholm to Rio: Summary Report of the Seminar Convened by the Canadian Department of External Affairs and International Tarade and the Department of Environment*, Willson House, Meech Lake, Quebec, December 8–9, 1991.

National Geographic. *Earth'88: Changing Geographic Perspectives – Proceedings of the Centennial Symposium* (Washington, DC: National Geographic Society, 1988).

National Science Foundation. "Scientists Closer to Identifying Cause of Antarctic Ozone Depletion," *National Science Foundation News* (October 20, 1986).

Neustadt, Richard E. *Alliance Politics* (New York: Columbia University Press, 1970).

North, Douglass C. *Institutions, Institutional Change and Economic Performance* (New York: Cambridge University Press,1990).

Nye, Joseph S. "The Other Global Power Shift," *Project Syndicate* (August 6, 2020). www.project-syndicate.org/commentary/new-technology-threats-to-us-national-security-by-joseph-s-nye-2020-08?utm_source=Project%20Syndicate%20Newsletter&utm_campaign=635fee703d-sunday_newsletter_09_08_2020&utm_medium=email&utm_term=0_

73bad5b7d8-635fee703d-105615321&mc_cid=635fee703d&mc_eid=6f34189ace&barrier=accesspaylog (accessed December 26, 2021).

O'Neill, Kate. "From Stockholm to Johannesburg and Beyond: The Evolving Meta-Regime for Global Environmental Governance," Paper prepared for presentation at the *2007 Amsterdam Conference on the Human Dimensions of Global Environmental Change* (May 24–26 2007). . http://citeseerx.ist.psu.edu/viewdoc/download?doi=10.1.1.662.8974&rep=rep1&type=pdf (accessed December 26, 2021).

Organization for Economic Co-operation and Development. *Environment and Economics: Results of the International Conference on Environment and Economics – 18th–21st June 1984* (Paris: OECD Publication, 1985).

Benefits Estimates and Environmental Decision-Making (Paris: OECD, 1992).

Oye, Kenneth A. "Explaining Cooperation under Anarchy: Hypotheses and Strategies," *World Politics*, vol. 38, no. 1 (October 1985), pp. 1–24.

Pallemaerts, Marc. "International Environmental Law from Stockholm to Rio: Back to the Future?," in Philippe Sands (ed.), *Greening International Law* (New York: The Free Press, 1994), pp. 1–19.

Parkinson, C. N. *Parkinson: The Law* (Boston: Houghton Mifflin, 1980).

Patrick, Stewart. "The World Order Is Starting to Crack," *Foreign Policy* (July 25, 2018). https://foreignpolicy.com/2018/07/25/the-world-order-is-starting-to-crack/ (accessed October 25, 2019).

"When the System Fails," *Foreign Affairs*, vol. 99, no. 4 (July/August 2020), pp. 40–50.

Pearce, Fred. "Beyond Rio's Disappointment: Finding a Path to the Future," *Yale Environment 360* (2012). https://e360.yale.edu/features/beyond_rios_disappointment_finding_a_path_to_the_future (accessed July 29, 2021).

Peters, B. Guy, Jon Pierre, and Desmond S. King, "The Politics of Path Dependency: Political Conflict in Historical Institutionalism," *The Journal of Politics*, vol. 67, no. 4 (November 2005), pp. 1275–1300.

Piccioni, Luigi. "Only One Earth: The Holy See and Ecology," in Wolfram Kaiser and John-Henrik Meyer (eds.), *International Organizations and Environmental Protection in the Global Twentieth Century: Actors, Agenda-Setting and Institution-Building* (New York: Berghahn Books, 2016), pp. 128–152.

Pierson, Paul. "Increasing Returns, Path Dependence, and the Study of Politics," *American Political Science Review*, vol. 94, no. 2 (June 2000), pp. 251–267.

Politics in Time: History, Institutions, and Social Analysis (Princeton, NJ: Princeton University Press, 2004).

Pierson, Paul and Theda Skocpol, "Historical Institutionalism in Contemporary Political Science," in Ira Katznelson and Helen V. Milner (eds.), *Political Science: State of the Discipline* (New York: W. W. Norton, 2002), pp. 693–721.

Powell, W. and Paul J. DiMaggio (ed.), *The New Institutionalism in Organizational Analysis* (Chicago: University of Chicago Press, 1991).

Pruitt, Dean G. "Ripeness Theory and the Oslo Talks," *International Negotiation*, vol. 2 (1997), pp. 237–250.

Putnam, Robert D. "Diplomacy and Domestic Politics," *International Organization*, vol. 42, no. 3 (Summer 1988), pp. 427–460.

Bowling Alone: The Collapse and Revival of American Community (New York: Simon & Schuster, 2000).

Raiffa, Howard. *The Art and Science of Negotiation* (Cambridge, MA: Belknap Press of Harvard University Press, 1982).

Raiffa, Howard, with John Richardson and David Metcalfe. *Negotiation Analysis: The Science and Art of Collaborative Decision Making* (Cambridge, MA: Belknap Press of Harvard University Press, 2002).

Ramphal, Shridath. *Our Country, The Earth* (New York: Island Press, 1992).

Glimpses of a Global Life (Toronto: Dundurn, 2014).

Rangarajan, Mahesh. "Striving for a Balance: Nature, Power, Science and India's Indira Gandhi, 1917–1984," *Conservation & Society*, vol. 7, no. 4 (2009) pp. 299–312.

Riker, William H. *The Theory of Political Coalitions* (New Haven, CT: Yale University Press, 1962).

Risse-Kappen, Thomas. "Ideas Do Not Float Freely: Transnational Coalitions, Domestic Structures and the End of the Cold War," *International Organization*, vol. 48, no. 2 (Spring 1994), pp. 185–214.

Rixen, Thomas, Lora Anne Viola, and Michael Zürn, "Historical Institutionalism and International Relations: Explaining Institutional Development in World Politics," in Thomas Rixen and Lora Anne Viola (eds.), *Historical Institutionalism and International Relations* (Oxford: Oxford University Press, 2016a), pp. 3–37.

Rixen, Thomas, Lora Anne Viola, and Michael Zürn, (eds.), *Historical Institutionalism and International Relations: Explaining Institutional Development in World Politics* (Oxford: Oxford University Press, 2016b).

Rocard, Michel. "Don't Blame It on Rio," *Project Syndicate* (February 22, 2012). www.project-syndicate.org/commentary/don-t-blame-it-on-rio?barrier=accesspaylog (accessed December 28, 2020).

Rojas-Suarez, Liliana. "From the Debt Crisis toward Economic Stability: An Analysis of the Consistency of Macroeconomic Policies in Mexico," International Monetary Fund Working Paper (March 1992), WP/92/L7.

Rousseau, Jean-Jacques. *A Discourse on Inequality* (London: Penguin Books, 1984).

Rowland, Wade. *The Plot to Save the World: The Life and Times of the Stockholm Conference on the Human Environment* (Toronto: Clarke, Irwin & Company Ltd., 1973).

Runnalls, David. "What Should Be Said at UNCED? Institutional Choices for the Rio Conference," in Department of External Affairs and International Trade and the Department of Environment, Canada, *From Stockholm to Rio, 1972–1992* (1991).

Sandholm, Thomas and Nir Vulkan, "Bargaining with Deadlines," *Proceedings of the American Association of Artificial Intelligence* (1999). https://dl.acm.org/doi/10.5555/315149.315224 (accessed December 26, 2021), pp. 44–51.

Schelling, Thomas C. "The Strategy of Conflict Prospectus for a Reorientation of Game Theory," *Journal of Conflict Resolution*, vol. 2, no. 3 (September 1958), pp. 203–264.

The Strategy of Conflict (Cambridge, MA: Harvard University Press, 1960).

"Hockey Helmets, Concealed Weapons, and Daylight Saving: A Study of Binary Choices with Externalities," *Journal of Conflict Resolution*, vol. 17, no. 3 (September 1973), pp. 381–428.

Schiavo-Campo, Salvatore and Hans W. Singer. *Perspectives of Economic Development* (Boston: Houghton Mifflin, 1970).

Schlesinger, Stephen. *Act of Creation: The Founding of the United Nations* (New York: Basic Books, 2004).

Scott, John. *Social Network Analysis*, 4th ed. (London: Sage, 2017).

Scott, W. Richard. "Institutions and Organizations: Towards a Theoretical Synthesis," in W. Richard Scott and John W. Meyer (eds.), *Institutional Environments and Organizations: Structural Complexity and Individualism* (Thousand Oakes, CA: Sage, 1994), pp. 55–80.

Sebenius, James K. "International Negotiation Analysis," in Victor A. Kremenyuk (ed.), *International Negotiation: Analysis, Approaches, Issues* (San Francisco, CA: Jossey-Bass, 2002), pp. 229–255.

Sebenius, James K., R. Nicholas Burns, and Robert H. Mnookin, *Kissinger the Negotiator: Lessons from Dealmaking at the Highest Level* (New York: Harper, 2018).

Seib, Gerald F. "In Crisis, Opportunity for Barack Obama," *Wall Street Journal* (November 21, 2008), p. A2.

Shabecoff, Philip. "U.S. Goes to Ecology Parley under Cloud of Doubt," *The New York Times* (May 5, 1982a), p. A2.

"U.S. Tells Ecologists to Trust Free Enterprise," *The New York Times* (May 12, 1982b), p. A7.

"The Environment Revisited," *The New York Times* (May 23 1982c), p. 183.

"Political and Economic Growth: A New Report Looks at the Links," *New York Times* (April 27, 1987a), p. B12.

"World Bank Offers Environmental Projects," *New York Times* (May 6, 1987b), p. A14.

A New Name for Peace: International Environmentalism, Sustainable Development, and Democracy (Hanover: University Press of New England, 1996).

Shepsle, Kenneth A. "Studying Institutions: Some Lessons from the Rational Choice Approach," *Journal of Theoretical Politics*, vol. 1, no. 2 (April 1989), pp. 131–147.

Sikkink, Kathryn. "Timing and Sequencing in International Politics," in: Orfeo Fioretos (ed.), *International Politics and Institutions in Time* (Oxford: Oxford University Press, 2017), pp. 231–250.

Simon, Hebert A. *Administrative Behavior: A Study of Decision-Making Processes in Administrative Organization* (New York: Macmillan, 1961).

Skocpol, Theda *States and Social Revolutions* (Cambridge: Cambridge University Press, 1979).

Skryms, Brian. "The Stag Hunt," Presidential Address: Pacific Division of the American Philosophical Association (March 2001). http://econ.ucsb.edu/~tedb/Evolution/skyrmsstaghunt.pdf (accessed July 31, 2020).

Slaughter, Anne-Marie. *The Chessboard and the Web: Strategies of Connection in a Networked World* (New Haven, CT: Yale University Press, 2017)

Sly, Eleanor. "Great Thunberg Criticises 'Empty Words' of World Leaders, Five Years since the Paris Agreement," *The Independent* (December 12, 2020). www.independent.co.uk/environment/climate-change/greta-thunberg-paris-agreement-video-b1770050.html (accessed January 5, 2021).

Smith, J. Eric. "The Role of Special Purpose and Nongovernmental Organizations in the Environmental Crisis," *International Organization*, vol. 26, no. 2 (Spring 1972), pp. 302–326.

Sohn, Louis B. "The Stockholm Declaration on the Human Environment," *Harvard International Law Journal*, vol. 14 (1973), pp. 423–515.

Soifer, Hillel David. "The Causal Logic of Critical Junctures," *Comparative Political Science*, vol. 45, no. 12 (2012), pp. 1572–1597.

Spector, Bertram I. "The Search for Flexibility on Financial Issues at UNCED: An Analysis of Preference Adjustment," in Bertram I. Spector, Gunnar Sjöstedt, and I. William Zartman (eds.), *Negotiating International Regimes: Lessons Learned from the United Nations*

Conference on Environment and Development (UNCED) (London: Graham & Trotman, 1994).

Speth, James Gustave and Peter M. Haas. *Global Environmental Governance* (Washington, DC: Island Press, 2006).

Stakeholder Forum for a Sustainable Future. *Sustainable Development in the 21st century (SD21): Review of Implementation of Agenda 21* (January 2012). https://sustainabledevelopment.un.org/content/documents/1126SD21%20Agenda21_new.pdf (accessed December 28, 2020).

Stein, Howard and E. Wayne Nafziger. "Structural Adjustment, Human Needs, and the World Bank Agenda," *The Journal of Modern African Studies*, vol. 29, no. 1 (1991), pp. 173–189.

Stone, Peter B. *Did We Save the Earth at Stockholm* (London: Earth Island, 1973).

Strandaneas, Jan-Gustav. "A Council for Sustainable Development: A Possible Outcome of the Rio+20 Process," Stakeholder Forum (2012).

Strong, Maurice. "The Stockholm Conference: Where Science and Politics Meet," *Ambio*, vol. 1, no. 3 (June 1972), pp. 73–78.

Where on Earth Are We Going? (Toronto: Vintage Canada, 2001).

Struthers, David. "The United Nations Environment Programme after a Decade: The Nairobi Session of a Special Character, May 1981," *Denver Journal of International Law & Policy*, vol. 12, no. 2 (1983).

Susskind, Lawrence E. *Environmental Diplomacy: Negotiating More Effective Global Agreements* (New York: Oxford University Press, 1994).

Tasker, John Paul. "Trudeau Unveils New Net-Zero Emissions Plan to Meet Climate Change Targets," *Canadian Broadcasting Corporation* (November 19, 2020). www.cbc.ca/news/politics/net-zero-emissions-1.5807877 (accessed January 5, 2021).

Thacher, Peter S. "Background on Institutional Options for Management of the Global Environment and Commons," in Environment Canada (ed.), *From Stockholm to Rio: Summary Report of the Seminar Convened by the Canadian Department of External Affairs and International Tarade and the Department of Environment,* Willson House, Meech Lake, Quebec, December 8–9, 1991.

Thelen, Kathleen. "How Institutions Evolve: Insights from Comparative-Historical Analysis," in James Mahoney and Dietrich Rueschemeyer (eds.) *Comparative Historical Analysis in the Social Sciences* (New York: Cambridge University Press, 2003), pp. 208–240.

Time. "America's Sewage System and the Price of Optimism," *Time* (August 1, 1969).

"Environment: Richard Nixon's New Issue," *Time*, special edition (February 2, 1970).

"Planet of the Year: Endangered Earth," *Time*, vol. 133, no. 1 (January 2, 1989).

Tolba, Mostafa K. "UNEP: Retrospect and Prospects: Introductory Statement to the Eighth Session of the Governing Council of UNEP," in Mostafa Kamal Tolba (ed.), *Development without Destruction: Evolving Environmental Perceptions* (Dublin: Tycooly International Publishing, 1982a).

"A Simple Choice – Co-operation or Catastrophe," *Uniterra*, vol. 2 (1982b), p. 6.

"The United Nations Conference on Desertification: A Review," *Mazingira*, vol. 6, no. 1 (1982c), pp. 14–16.

"Environmental Protection: Address to the Japan Environmental Agency Symposium, Tokyo," in Mostafa Kamal Tolba (ed.), *Development without Destruction: Evolving Environmental Perceptions* (Dublin: Tycooly International Publishing, January 1982d).

Tolba, Mostafa K. "Environmental Protection: Address to the Japan Environmental Agency Symposium, Tokyo," in Mostafa Kamal Tolba (ed.). *Development without Destruction: Evolving Environmental Perceptions* (Dublin: Tycooly International Publishing, 1982e).

(ed.). *Sustainable Development: Constraints and Opportunities* (London: Butterworths, 1987).

"Environmental Protection: Address to the Japan Environmental Agency Symposium, Tokyo," in Mostafa Kamal Tolba *Saving Our Planet: Challenges and Hopes* (London: Springer, 1992).

Global Environmental Diplomacy: Negotiating Environmental Agreements for the World, 1973–1992 (Cambridge: MIT Press, 1998).

Tolba, Mostafa Kamal and Asit K. Biswas (eds.). *Earth and Us: Population-Resources-Environment-Development* (Oxford: Butterworth-Heinemann Ltd., 1991).

Tolba, Mostafa Kamal, Osama A. El-Kholy, Essam E. Hinnawi, M. W. Holdgate, D. F. Michael, and R. E. Munn, *The World Environment: 1972–1992: Two Decades of Challenge* (London: Chapman & Hall, 1992).

Tomlin, Brian, Norman Hillmer, and Fen Hampson, *Canada's International Policies: Agendas, Alternatives, and Politics* (Toronto: Oxford University Press, 2007).

Tower, Samuel A. "Stamps; U.N. Commemoratives," *New York Times*, Arts and Leisure 2 (March 7, 1982).

UNESCO (Abdellatif Benachenhou). *Environment and Development: Problems and Prospects for Sustainable Development* (Paris: UNESCO, 1992).

United Kingdom Department of Environment. *Our Common Future: A Perspective by the United Kingdom on the Report of the World*

Commission on Environment and Development (London: United Kingdom Government, 1988).

United Nations Conference on Trade and Development. *Technology Transfer and Development in a Changing International Environment: Policy Challenges and Options for Cooperation* (New York: United Nations, 1992).

United Nations Environment Programme. "Report of the Governing Council (Session of a Special Character and Tenth Session)" (1982). https://wedocs.unep.org/bitstream/handle/20.500.11822/17274/82_05_GC10_%20special_character_report_of_the_GC_10_1982.pdf?sequence=10&isAllowed=y (accessed December 26, 2021).

The State of the Environment 1972–1982 (Nairobi: Kanaya Designs, 1982).

"Malmö Ministerial Declaration," Report of the Governing Council, 6th Special Session (May 29–31, 2000), A/55/25, UNEP/GC/DEC/SS.V1/1.

Consultative Group of Ministers or High-Level Representatives on International Environmental Governance/Nairobi-Helsinki Outcome (December 13, 2010), UNEP/GC.26/18.

"Letter from the Executive Director" (2019) Annual Report., www.unep.org/resources/annual-report/letter-executive-director-2019-review (accessed December 26, 2021).

"Making Peace with Nature: A Scientific Blueprint to Tackle the Climate, Biodiversity and Pollution Emergencies" (February 2021). https://wedocs.unep.org/xmlui/bitstream/handle/20.500.11822/34948/MPN.pdf (accessed July 17, 2021).

United Nations Environment Programme (Committee of International Development Institutions on the Environment). *Workshop on Environmental and Natural Resource Accounting: Nairobi, 24–26 February 1992: Summary Record* (Nairobi: UNEP, 1992).

United Nations Environment Programme (Hussein Abaza). *Consultative Meeting on Environment and Economics, Nairobi, 26–28 February 1992: Summary Record* (Nairobi: UNEP, 1992).

United Nations Environment Programme and World Health Organization. *Urban Air Pollution in Megacities of the World* (Oxford: Blackwell Press, 1992).

Vabulas, Felicity and Duncan Snidal. "Organization without Delegation: Informal Intergovernmental Organizations and the Spectrum of Intergovernmental Arrangements," *Review of International Organizations*, vol. 8, no. 2 (2013), pp. 193–220.

Wagner, Travis. "Hazardous Waste: Evolution of a National Environmental Problem," *Journal of Policy History*, vol. 16, no. 4 (2004), pp. 306–331.

Wapner, Paul Kevin. "World Summit on Sustainable Development: Toward a Post-Jo'Burg Environmentalism," *Global Environmental Politics*, vol. 3, no. 1 (February 2003), pp. 1–10.

Ward, Barbara. *Spaceship Earth* (New York: Columbia University Press, 1966).

"Eating Seed Corn, Burning Dung," *The Observer* (June 20, 1982), p. 26.

Ward, Barbara and René Dubos. *Only One Earth: The Care and Maintenance of a Small Planet* (New York: W. W. Norton, 1972).

Ward, Barbara, René Dubos, Thor Heyerdahl et al. *Who Speaks for the Earth? Seven Citizens of the World on Major Issues of the Global Environment* (New York: W. W. Norton, 1981).

Weibust, Inger. *Green Leviathan: The Case for a Federal Role in Environmental Policy* (Farnham: Routledge, 2009).

Weingast, Barry R. "Political Institutions: Rational Choice Perspectives," in Robert Goodin and Hans-Dieter Klingemann (eds.), *A New Handbook of Political Science* (Oxford: Oxford University Press, 1996), pp. 167–190.

Welch, David A. *Painful Choices: A Theory of Foreign Policy Change* (Princeton, NJ: Princeton University Press, 2005).

Wendt, Alexander. "The Agent-Structure Problem in International Relations Theory," *International Organization*, vol. 41, no. 3 (Summer 1987), pp. 335–370.

"Anarchy Is What States Make of It: The Social Construction of Power Politics, *International Organization*, vol. 46, no. 2 (Spring, 1992), pp. 391–425.

Werksman, Jacob. "The Conference of the Parties to Environmental Treaties," in Jacob Werksman (ed.), *Greening International Institutions* (London: Earthscan, 1996), pp. 55–68.

Widmaier, Wesley W., Mark Blyth, and Leonard Seabrooke, "Symposium on the Social: Construction of Wars and Crises as Openings for Change: Exogenous Shocks or Endogenous Constructions? The Meanings of Wars and Crises, *International Studies Quarterly*, vol. 51, no. 4 (December 2007), pp. 747–759.

Wijkman, Anders. *Natural Disasters: Acts of God or Acts of Man* (London: International Institute for Environment and Development, 1984).

Williams, Marc. "Re-Articulating the Third World Coalition: The Role of the Environmental Agenda," *Third World Quarterly*, vol. 14, no. 1 (1993), pp. 7–29.

Williamson, Oliver E. "The New Institutional Economics: Taking Stock, Looking Ahead," *Journal of Economic Literature*, vol. 38, no. 3 (September 2000), pp. 595–613.

World Bank (Patrick Low). *International Trade and the Environment* (Washington, DC: World Bank, 1992).

World Commission on Environment and Development (hereafter WCED). *Our Common Future* (Oxford: Oxford University Press, 1987).

World Health Organization. *Our Planet, Our Health: Report of the WHO Commission on Health and Environment* (Geneva: WHO, 1992a).

WHO Commission on Health and Environment: Report of the Panel on Energy, WHO/EHE/92.3 (Geneva: WHO, 1992b)

WHO Commission on Health and Environment: Report of the Panel on Food and Agriculture, WHO/EHE/92.2 (Geneva: WHO, 1992c)

WHO Commission on Health and Environment: Report of the Panel on Urbanization, WHO/EHE/92.5 (Geneva: WHO, 1992d).

World Resources Institute, The World Conservation Union, United Nations Environment Programme; in consultation with Food and Agriculture Organization, United Nations Education, Scientific and Cultural Organization. "Global Biodiversity Strategy: Guidelines for Action to Save, Study, and Use Earth's Biotic Wealth Sustainably and Equitably" (Washington, DC: World Resource Institute, 1992).

Young, Oran R. "Regime Dynamics: The Rise and Fall of International Regimes," *International Organization*, vol. 36, no. 2 (Spring 1982), pp. 277–297.

"The Politics of International Regime Formation: Managing Natural Resources and the Environment," *International Organization*, vol. 43, no. 3 (Summer 1989), pp. 349–375.

Institutional Dynamics: Emergent Patterns in International Environmental Governance (Cambridge: MIT Press, 2010).

Zartman, I. William. "Bargaining and Conflict Resolution," in Edward A. Kolodziej and Roger E. Kanet (eds.), *Coping with Conflict after the Cold War* (Baltimore: Johns Hopkins University Press, 1996).

Zartman, I. William and Maureen Berman. *The Practical Negotiator* (New Haven, CT: Yale University Press, 1982).

Zelli, Fariborz. "The Fragmentation of the Global Climate Governance Architecture," *WIREs Climate Change*, vol. 2 (March/April 2011), pp. 255–270.

Zucker, Lynne. "The Role of Institutionalization in Cultural Persistence," in Walter W. Powell and Paul J. DiMaggio (eds.), *The New Institutionalism in Organizational Analysis* (Chicago: University of Chicago Press, 1991), pp. 83–101.

Zürn, Michael. "Historical Institutionalism and International Relations – Strange Bedfellows?" in Thomas Rixen, Lora Anne Viola, and Michael Zürn (eds.), *Historical Institutionalism and International Relations: Explaining Institutional Development in World Politics* (Oxford: Oxford University Press, 2016), pp. 199–228.

Index

CPSIA information can be obtained
at www.ICGtesting.com
Printed in the USA
LVHW081917050622
720533LV00005B/237

9 781009 165884